A FIRE INVESTIGATOR'S HANDBOOK

Wayne P. Petrovich began his career in 1972 with the City of Hollywood Florida Fire Department as a firefighter. In 1979, he joined the fire department's Bureau of Fire Prevention as an Inspector/Investigator.

Petrovich left the fire department in 1981 and joined the Florida State Fire Marshal's Office as a law enforcement investigator with the Bureau of Fire/Arson Investigations, assigned to the West Palm Beach Office.

Investigator Petrovich earned an AA degree in Criminal Justice, and has attended many schools and seminars in the field of fire investigations, including courses at the National Fire Academy, FBI Academy, and the Bureau of Alcohol, Tobacco and Firearm's Arson for Profit School.

As a senior fire investigator with the State Fire Marshal's Office, Petrovich has testified in numerous civil and criminal court cases both in the state and federal judicial systems. He is currently one of the instructors at the State Fire College in Ocala and for the Indian River Community College Fire Science Program. Investigator Petrovich has been a guest speaker for many law enforcement and fire service departments and seminars throughout the country.

A FIRE INVESTIGATOR'S HANDBOOK

Technical Skills for Entering, Documenting and Testifying in a Fire Scene Investigation

By
WAYNE P. PETROVICH

With a Foreword by
Guy E. Burnette, Jr.

CHARLES C THOMAS • PUBLISHER, LTD.
Springfield • Illinois • U.S.A

Published and Distributed Throughout the World by

CHARLES C THOMAS · PUBLISHER, LTD.
2600 South First Street
Springfield, Illinois 62794-9265

©*1998 by* CHARLES C THOMAS · PUBLISHER, LTD.
ISBN 0-398-06794-5 (cloth)
ISBN 0-398-06795-3 (paper)
Library of Congress Catalog Card Number: 97-14973

With THOMAS BOOKS *careful attention is given to all details of manufacturing
and design. It is the Publisher's desire to present books that are satisfactory as to their
physical qualities and artistic possibilities and appropriate for their particular use.*
THOMAS BOOKS *will be true to those laws of quality that assure a good name
and good will.*

Printed in the United States of America
CR-R-3

Library of Congress Cataloging in Publication Data

Petrovich, Wayne P.
    A fire investigator's handbook : technical skills for entering,
documenting, and testifying in a fire scene investigation / by
Wayne P. Petrovich ; with a foreword by Guy W. Burnette, Jr.
        p.   cm.
    Includes bibliographical references and index.
    ISBN 0-398-06794-5 (cloth). – ISBN 0-398-06795-3 (pbk.)
    1. Fire investigation.   I. Title.
TH9180.P384   1997
363.37'65–dc21   97-14973
                                                        CIP

This book is dedicated to the families of fire investigators and especially to my wife, Debra, who spend countless hours, alone, at any given moment, day or night, as we pursue the fire's origin and cause.

FOREWORD

The crime of arson remains the single least-often and least-effectively prosecuted crime in America. It is the nature of arson that it is usually committed under circumstances where there are no eyewitnesses to the offense and much of the physical evidence is destroyed in the fire. It is left to the investigator to piece together the evidence and direct the latent investigation necessary to effect an arrest. Arson is inevitably a circumstantial crime requiring thorough investigation and a methodical approach to fit together the pieces of the puzzle.

The proper collection of physical evidence and the effective use of demonstrative evidence at trial are the cornerstones of an arson case. With this text, Wayne Petrovich has provided a powerful tool for the fire investigator dedicated to doing the job right. The experiences and lessons learned from the thousands of cases he has worked are brought together in this unique book for every fire investigator to learn and apply. This represents an important step forward in the fight against arson. The battle lines have been drawn in this war and the weapons are here for those who will take up the call to arms. It is a war we cannot afford to lose.

Guy E. "Sandy" Burnette, Jr.

PREFACE

This book is written for two types of fire investigators:. the novice, who is seeking to develop skills in fire investigation, and the veteran, who wants to enhance his or her fire investigative techniques.

To assist both types of investigators, this book is divided into three sections. The first section advises the investigator on how to legally enter the fire scene. The second section explains the methods of documenting a fire scene, and the third section describes the procedures of becoming a good witness while testifying in court.

The book stresses that each section has an important relationship to the other. To omit techniques and/or to be not properly prepared could lead to damaging testimony for the fire investigator.

ACKNOWLEDGMENTS

Without the assistance of the following individuals this book could not have been written. A large debt of gratitude is given to Ann Kaklamanos, librarian for the Law Library of the Florida Fourth District Court of Appeals, who helped in my research through countless volumes of law books and periodicals, and to Marge Fickley, who gave up hours of free time to type and proofread the manuscript. A special thanks to Detective Dennis Saling of the Palm Beach County Sheriff's Office, Bomb and Arson Squad, for his assistance in photography. To attorney Sandy "Guy" Burnette Jr. for his advice and foreword, and to Collin Holmes and Michele Petrovich for their illustrations. Lastly, I would like to thank all of the fire investigators and law enforcement personnel who taught me, through the ashes of their investigations, what it means to be a creditable fire investigator.

INTRODUCTION

Throughout my career as a fire investigator, many hours have been spent with other investigators, exchanging views on the education and skills obtained in our line of work. Many agree that the basic knowledge of fire investigation found in books, schools and seminars is a good foundation for the beginner investigator as well as a refresher for the seasonal investigator.

As the fire investigator becomes experienced, the range of knowledge becomes sophisticated. Not only does the investigator learn fire behavior and basic fire scene investigation techniques, he or she must also become proficient in advanced methods and problems such as criminal law, the proper techniques for documenting the fire scene, courtroom demeanor, and expert testimony, just to name a few, that may be necessary to successfully conclude the investigation.

During my twenty-two years as both a fire and law enforcement official in the field of fire investigations, I have constantly seen the need for the development of educational schools and seminars for the advanced investigator. Through these educational channels, information not readily accessible can be disseminated. The move for higher learning must continue for both the beginner and advanced fire investigator if we, as fire investigators, are to keep up with the rapid progress of science and industrial technology being introduced into today's everyday life.

Hopefully, this book will become one of many written by fire investigators that will give support to the investigator in his or her quest to document the fire scene.

CONTENTS

Page

Foreword by Guy E. Burnette, Jr.   vii
Preface   ix
Introduction   xiii

SECTION 1 - ENTERING THE FIRE SCENE LEGALLY

Introduction   5
Chapter 1   EXIGENT OR EMERGENCY CIRCUMSTANCE   9

    Michigan v. Tyler   9
    Michigan v. Clifford   15
    Administrative Warrant   19

Chapter 2   CONSENT   29

    Who Has the Right to Give Consent   31
    Consent Using Accelerant Detection Canines   34

Chapter 3   THE PLAIN VIEW DOCTRINE   41

Chapter 4   ABANDONMENT   51

Chapter 5   OPEN FIELD   57

Chapter 6   MIRANDA WARNINGS   65

SECTION 2 - DOCUMENTING THE FIRE SCENE

Chapter 7   DEMONSTRATIVE EVIDENCE   83

Chapter 8   SKETCHES, DIAGRAMS, MAPS, ILLUSTRATIONS   89

    Line Weights of a Sketch   93
    Types of Sketches   99
    Rough Sketches   104
    Types of Measurements   105

Tools and Materials for the Rough Sketch   111
Rules for Drawing the Rough Sketch   116
The Finished Sketch   116
The Computer Sketch   119

Chapter 9   PHOTOGRAPHS   125

The Modern Camera   128
Types of Camera   132
Film   134
Documentation of the Fire Scene   136
Photographing Evidence   154
Photographing a Fire Fatality   154
Aerial Photographs   155
Lighting   155
Maintenance of the Camera   162
Admissibility of Photographs   164

Chapter 10   VIDEO   169

History of Video   170
Types of Video Camera   171
Lighting   173
Sound   176
Composition   179
Special Effects   185
The Fire Scene Production   189
Video in the Legal System   191

Chapter 11   FIRE SCENE MODELS   203

Admissibility of the Fire Model in Court   204
Types of Fire Scene Models   209
Construction of the Model   227
Materials for Model Construction   231
Building the Model in Two Parts   239

Chapter 12   COMPUTERS AND THE FIRE INVESTIGATION   249

Computer Animation   252

Computer Simulation 253
Admissibility of Animation and Simulation 255
The Future 261

SECTION 3 - COURTROOM PRESENTATION

Chapter 13   KNOWING THE LEGAL SYSTEM 269

Types of Judicial Systems 269
Sequence of Events in the Courtroom 273

Chapter 14   TESTIFYING AS A WITNESS 285

Deciding What to Wear 287
Day of Court Appearance 288
Helpful Body Language Tips 291
More Information on the Cross Examination 297

Chapter 15   THE EXPERT WITNESS 301

Qualifying as an Expert 303

Index 323

A FIRE INVESTIGATOR'S HANDBOOK

SECTION 1

ENTERING THE FIRE SCENE LEGALLY

FOURTH AMENDMENT RIGHTS

EXCEPTIONS TO THE SEARCH WARRANT

EXIGENT OR EMERGENCY CIRCUMSTANCES
MICHIGAN V. TYLER
MICHIGAN V. CLIFFORD
ADMINISTRATIVE WARRANT

CONSENT TO SEARCH

PLAIN VIEW DOCTRINE

ABANDONMENT

OPEN FIELD

MIRANDA RIGHTS

INTRODUCTION

In legal terms, "the fruit of the poison tree doctrine" deals with the illegal conduct of an official during a search and seizure. The official, to name a few, may be a law enforcement officer, building inspector or a fire investigator. The conduct of the official may be conscious or unconscious. Either way, if the evidence is seized when a search is conducted while in violation of an accused person's Fourth Amendment rights, then that evidence cannot be used against said individual in a criminal trial.

The phrase "fruit of the poison tree" is derived from the belief that once the tree is poisoned, then it must be assumed that so too is the tree's fruit. In law, the "fruit of the poison tree" would be the primary evidence initially seized illegally, and any other evidence that has been tainted because of its relationship to the illegal activity. For example, when a confession is made, in part, by showing the suspect the evidence obtained during an illegal search, that confession and any evidence seized during the search may be found inadmissible in a criminal trial.

There are two landmark cases involving evidence seized during an illegal search. In 1914, the United States Supreme Court, in the case *Weeks v. US* (232 US 383, 34 S Ct 341, 58 L ed 652), established the doctrine called "the fruit of the poison tree" or the Exclusionary Rule. The court further stated that the rule would apply to all federal prosecutors.

In 1961, the United States Supreme Court, in the case *Mapp v. Ohio* (367 US 643, 81 S Ct 1684, 6 L ed 2d 1081) stated that under the Fourth and Fourteenth Amendments, any evidence seized through an illegal search and seizure was inadmissible in state as well as in federal courts.

For this reason, the legal aspects of how a fire investigator should enter a fire scene are discussed first. For, if the fire investigator does not know his or her legal limitations and the rights given to individuals through the United States Constitution, that investigator may find his evidence inadmissible and countless hours of hard work thrown out.

FOURTH AMENDMENT OF THE UNITED STATES CONSTITUTION

As representatives of government agencies, law enforcement and fire personnel should fully understand how to legally conduct a fire scene investigation. Before entering a fire scene, the investigator needs to be aware that a person is protected from unreasonable search and seizure through the Fourth Amendment of the United States Constitution. The Fourth Amendment states:

> The right of the people to be secured in their persons, houses, papers and effects, against unreasonable searches and seizures, shall not be violated and no warrants, shall be issued, but upon probable cause, supported by oath or affirmation, and particularly describing the place to be searched, and the person or things to be seized.

What the Fourth Amendment states is that in order to have a legal search and seizure of evidence, there must be a warrant based on probable cause and issued by a judicial officer. This relieves the law enforcement officer involved in the investigation from a biased decision to search and seize, thus bringing the probable cause to a neutral party, the judicial judge. Therefore, any search and seizure of evidence made without a warrant is considered unreasonable and in violation of the Fourth Amendment.

To understand what a search does, a definition from the book, *Search and Seizure, A Treatise on the Fourth Amendment* 2nd edition by Wayne R. LaFave, best describes what constitutes a search:

Some exploratory investigation, or an invasion and quest, a looking for or seeking out. The quest may be secret, intrusive, or accomplished by force, and it has been held that a search implies some sort of force, either actual or constructive, much or little. A search implies a prying into hidden places for that which is concealed and that the object searched for has been hidden or intentionally put out of the way. While it has been said that ordinarily searching is a function of sight it is generally held that the mere looking at that which is open to view is not a "search."

Another excellent definition comes from the *Black's Law Dictionary*, 4th Edition, which states a search as:

> An examination of a man's house, or other buildings or premises, or of his person, with a view to the discovery of contraband or illicit or stolen   property,

or some evidence of guilt to be used in the prosecution of a criminal action for some crime or offense with which he is charged.

Another term that needs to be clarified in the Fourth Amendment is the word "houses." For the Fourth Amendment interpretation, a "house" involves two different classes. One class engages where a person lives, either in a whole structure or a part of a structure that can be owned, rented, or leased, either permanently or temporarily. Family dwellings, apartments, individual townhouses, rooming houses, hotels/motels, etc. are some examples.

The second class is a structure, either whole or in part, where an individual works. This area also can be owned, rented or leased, permanently or temporarily. The work area can be as small as a work desk, a one-room business, an office, a warehouse, or even as large as an industrial complex.

There are many state and federal court cases involving decisions concerning search and seizure. One of the landmark cases on this subject is United States Supreme Court case *Katz v. United States* (389 US 347, 19 L ed 2d 576, 88 S Ct 507). This decision looks at the constitutionality of the Fourth Amendment. In the opinion for the court, delivered by Justice Stewart, he stated:

> .....once it is recognized that the Fourth Amendment protects people– and not simply "areas"–against unreasonable searches and seizures, it becomes clear that the reach of that Amendment cannot turn upon the presence or absence of a physical intrusion into any given enclosure. (*Katz v.US* 389 US 353, 19 L ed 2d 583)

The court also stated that a person had the reasonable expectation of privacy. Justice Stewart wrote:

> What a person knowingly exposes to the public, even in his own home or office, is not a subject of Fourth Amendment protection... But, what he seeks to preserve as private, even in an area accessible to the public, may be constitutionally protected. (*Katz v US* 389 US 351, 19 L ed 2d 582)

The court also wanted to make clear that if a search is to be conducted, then it would be done so according to the requirements set forth in the Fourth Amendment. Otherwise, the search and seizure would be unreasonable, and thereby unlawful. Justice Stewart stated:

> Over and again this court has emphasized that the mandate of the [Fourth] Amendment requires adherence to judicial processes, *United States v. Jeffers* 342 US 48, 51 96 L ed 59, 64 72 S Ct 93, and that searches conducted outside the judicial process, without prior approval by judge or magistrate, are per se unreasonable under the Fourth Amendment –subject only to a few specifically established and well delineated exceptions. (*Katz v. US* 389 US 357, 19 L ed 2d 585)

The United States Supreme Court has also interpreted in the Fourth Amendment several exceptions to the warrant requirement. These exceptions of a warrantless search and seizure can be made under exigent or emergency circumstances, consent to search, in plain view, abandonment and in the open field. (These exceptions will involve the fire investigator or law enforcement officer at the fire scene and will be discussed in length.) Other exceptions are hot pursuit, incident to a lawful arrest, inventory search, stop and frisk, consent (custodial), border search, and searches involving trunks of vehicles.

Remember, that if a warrantless search is performed, it is in violation of the Fourth Amendment and up to the investigator to justify to the court as to why a warrant was not obtained before entry. The burden of proof is on the investigator and not on the neutral party who would have read the probable cause.

Chapter 1

EXIGENT OR EMERGENCY CIRCUMSTANCE

A search conducted under exigent or emergency circumstance deals with a warrantless search based on the fact that the time needed to obtain a warrant is not possible or practicable. The following are some examples of exigent or emergency circumstances. One, the hot pursuit, which allows law enforcement officers to follow suspects they are chasing to enter the same building as the fleeing suspect (U.S. Supreme Court case 1967 *Warden v. Hayden*, 387 US at 299 87 S Ct at 1646, 18 L ed 2d at 787). A second exigent circumstance is when a law enforcement officer has reason to believe a violent crime is being committed (*People v Clark*, 262 Cal. App. 2d 471, 476-477, 68 Cal. Rptr. 713, 717). Third, a law enforcement officer can enter a building without a warrant when he or she believes that evidence is going to be destroyed.

For firefighters and fire investigators, exigent or emergency circumstances involve a landmark decision in the United States Supreme Court entitled *Michigan v. Tyler* (436 US 499, 56 L Ed 2d 486) in 1978. This decision supports a warrantless entry by fire fighters for suppression activities and redefines the fire investigator's activities in the subsequent investigation as to the origin and cause of the fire. To fully understand the implications of *Michigan v. Tyler*, the facts of the case will be discussed as well as the opinion of the majority of the United States Supreme Court.

MICHIGAN V. TYLER
The Fire Investigation as to Origin and Cause

Just before midnight on January 21, 1970, fire personnel responded to a structure fire involving a furniture store leased by Loren Tyler who, along with a partner, conducted a business called Tyler's Auction. During the fire suppression activities, a fire lieutenant became aware of two plastic containers of a possible flammable liquid which were found in the building. The fire chief who was in charge of determining the fire cause and writing the fire report, arrived at the scene at approximately 2:00 AM on January 22, 1970. When the fire chief arrived, fire department personnel were involved in overhaul procedures. After a brief discussion with the fire lieutenant, they decided to enter the fire scene and observe the containers. Concluding the fire cause to be a possible arson, the fire chief notified the police detective who arrived at the scene at approximately 3:20 AM on January 22nd. The detective photographed the containers and parts of the interior until he had to leave the fire scene due to the smoke and steam. The containers were taken from the fire scene and taken into evidence by the police detective. At 4:00 AM, January 22nd, the fire was declared extinguished. At that time, fire and police personnel returned to quarters leaving the fire scene unattended.

After daylight, at approximately 8:00 AM, on January 22nd, four hours after leaving, fire personnel returned to the fire scene for a cursory examination. On January 22nd at 9:00 AM, the police detective returned to the fire scene with an assistant fire chief. With the heat, smoke and darkness gone, a more thorough investigation was conducted. Evidence of a suspicious fire was observed in the form of burn marks in the carpet along with some other evidence found on the stairway. All evidence found was collected and taken without a search warrant or consent.

On February 16th, an investigator from the Michigan State Police, Arson Section, arrived at Tyler's Auction to conduct his investigation. Photographs were taken as well as several pieces of evidence. The state investigator returned the day after, four days later, seven days after the fire and 25 days after the fire. On each of these days, evidence was seized which indicated the fire was arson. Each entry into the building was made without a search warrant or consent.

At the trial, Tyler and his partner were found guilty through the testimony of all the individuals involved in the fire investigation and the physical evidence seized.

Tyler appealed his conviction to the Michigan Supreme Court. The Michigan Supreme Court stated that the containers found during the suppression were legally seized. However, once the fire fighters extinguished the fire and left the scene, all evidence seized during the warrantless search violated Mr. Tyler's Fourth Amendment rights. The United States Supreme Court agreed to hear the case when the state prosecutor appealed the Michigan Supreme Court's decision.

The United States Supreme Court Decision

The reason *Michigan v. Tyler* became the first landmark case concerning fire scene investigation was due to the court's definitive discussion on the relationship between the entry/search and the Fourth Amendment.

The United States Supreme Court agreed with the Michigan Supreme Court concerning the warrantless entry by the fire fighters to suppress the fire under exigent circumstances. The court also stated that if evidence of arson was observed during the suppression activities that evidence could be seized. In the majority opinion of the United States Supreme Court, delivered by Justice Stewart, he stated:

> A burning building clearly presents an exigency of insufficient proportions to render a warrantless entry "reasonable." Indeed, it would defy reason to suppose that firemen must secure a warrant or consent before entering a burning structure to put out the blaze. And once in the building for this purpose, firefighters may seize evidence of arson that is in plain view (*Coolidge v. New Hampshire*, 403 US 443, 465-466). Thus, the Fourth and Fourteenth Amendments were not violated by the entry of the firemen to extinguish the fire at Tyler's Auction, nor by Fire Chief See's removal of the two plastic containers of flammable liquid found on the floor of one of the showrooms. (436 US 509, 56 L ed 2d 498)

The next issue addressed by the United States Supreme Court dealt with the determination of the fire cause. It is important for the fire personnel to determine the fire cause; one, for safety reasons and second, for the preservation of criminal evidence. Justice Stewart stated:

> Fire officials are charged not only with extinguishing the fires, but with finding their causes. Prompt determination of the fire's origin may be necessary to prevent its recurrence, as through the detection of continuing dangers such as faulty wiring or a defective furnace. Immediate investigation may also be necessary to preserve evidence from intentional or accidental destruction. And, of course, the sooner the officials complete their duties, the less will be their sub-

sequent interference with the privacy and recovery efforts of the victims. For these reasons, officials need no warrant to remain in a building for a reasonable time to investigate the cause of a blaze after it has been extinguished. And if the warrantless entry to put out the fire and determine the cause is constitutional, the warrantless seizure of evidence while inspecting the premises for these purposes also is constitutional. (436 US 510, 56 L ed 2d 498)

In the *Michigan v. Tyler* case, the United States Supreme Court stated that the entries made during the suppression were reasonable and the seizure of evidence during the suppression reasonable. The Court went further and stated that the warrantless entries and searches on the morning of January 22nd were a continuation of the initial entry that was halted at 4:00 AM on January 22nd. The Court concluded that the initial investigation could not properly be performed due to the darkness, steam and heat. The investigation did continue shortly after daylight, when the darkness, steam and heat were absent and within a "reasonable time" after the fire.

The one thing the United States Supreme Court failed to address was the definition of "reasonable time." If one is to read between the lines, "reasonable time" refers to the first opportune time the scene can be properly worked.

Once the exigent circumstance is over and the reasonable time has elapsed, then the fire investigator must either get a consent or a search warrant to enter onto the fire scene. This was addressed by the Supreme Court involving the entry and search of the Tyler Auction fire scene after January 22nd. The Court stated:

> That the search and seizure of evidence after January 22nd were....clearly detached from the initial exigent and warrantless entry. Since all of these searches were conducted without valid warrants and without consent, they were invalid under the Fourth and Fourteenth Amendments and any evidence obtained as a result of those entries must, therefore, be excluded at the respondent's [Tyler] retrial. (436 US 511, 56 L ed 2d 499)

The Supreme Court faced another problem. The Court felt the fire investigator needed to get onto the fire scene to make the determination as to the origin and cause of the fire, but when the reasonable time had elapsed and consent had not been given to gain entry, a warrant was needed. To obtain a search warrant, probable cause is required and since the fire investigator needs to be on the scene to make the determination, the fire for all practical purposes would have to be classified as undetermined. The classification of undetermined nullifies the probable cause requirement and the warrant. The investigator can develop probable cause of a suspicious or incendiary fire through witness accounts of what they heard or observed from the fire scene. This will enable the fire investigator to get onto the scene with a search warrant. To overcome this obstacle and still be within the scope of the Fourth Amendment, the Court felt this type of entry could be made pursuant to an administrative search warrant. The opinion of Justice Stewart:

> Official entries to investigate the cause of the fire must adhere to the warrant procedures of the Fourth Amendment. In the words of the Michigan Supreme Court: "Where the cause [of the fire] is undetermined, and the purpose of the investigation is to determine the cause and to prevent such fires from occurring or recurring, a...search may be conducted pursuant to a warrant issued in accordance with reasonable legislative or administrative standards. (436 US 508, 56 L ed 2d 497)

The administrative standards that the Michigan Supreme Court was alluring to was based on two United States Supreme Court cases, *See v. Seattle* (387 US 541, 18 L ed 2d 943) and *Camara v. Municipal Court* (387 US 523, 18 L ed 2d 930.) [Look under the topic Administrative Warrant for further discussion on this subject.]

The Court also made a distinction that the Fourth Amendment covers firefighters in their search for fires, as well as police officers who search a crime scene. Justice Stewart stated:

> There is no diminution in a person's reason able expectation of privacy nor in the protection if the Fourth Amendment simply because the official conducting the search wears the uniform of a firefighter rather than a policeman, or because his purpose is to ascertain the cause of the fire rather than to look for evidence of a crime, or because the fire might have been started deliberately. Searches for administrative purposes, like searches for evidence of a crime, are encompassed by the Fourth Amendment. (436 US 506, 56 L ed 2d 496)

Justice Stewart again reiterated the importance of any search within a private area when he stated a quote from the Camara v. Municipal Court of the City and County of San Francisco decision:

> And under that [Fourth] Amendment, one governing principle, justified by history and current experience, has consistently been followed: Except in certain fully defined classes of cases, a search of private property without proper con-

sent is "unreasonable" unless it has been authorized by a valid search warrant. (436 US 506, 56 L ed 2d 496)

The administrative warrant for the fire investigator is to show a neutral party, the magistrate, the purpose of the search, the number of times prior entries were made (if any), the scope of the search to be conducted, the time of day the search is to be conducted, the lapse of time the search will be conducted since the fire, the continued use of the building and is the building being secured by the owner. This type of warrant, as well as the criminal search warrant, is required to protect, in this case, the fire victim, from governmental harassment and by keeping the intrusion to a minimum.

The Supreme Court concluded that the entry made by the fire chief and his assistants during the early morning of January 22nd did not need a warrant due to the exigent circumstance exception. When the fire chief returned, after the "darkness, steam and heat" disappeared, the investigation was a continuation of the early morning entry and therefore did not need a warrant. However, all other entries after January 22nd were without consent or a warrant, making the entries a violation of the Fourth Amendment. With the entries unlawful, the evidence also became unlawful and could not be used in the retrial.

In conclusion, the Supreme Court made a summation as to how fire investigators should conduct their entries and investigation. The court's opinion delivered by Justice Stewart stated:

> We hold that an entry to fight a fire requires no warrant, and that once in the building, officials may remain there for a reasonable time to investigate the cause of the blaze. Therefore, additional entries to investigate the cause of the fire must be pursuant to the warrant procedures governing administrative searches. See Camara, supra, at 534-539, *See v. City of Seattle*, supra, at 544-545; *Marshal v. Barlow's Inc.*, ante, at 320-321, at 56 L ed 2d 305, 12-13). Evidence of arson discovered in the course of such investigations is admissible at trial, but if the investigating officials find probable cause to believe that arson has occurred and require further access to gather evidence for a possible prosecution they may obtain a warrant only upon a traditional showing of probable cause applicable to searches for evidence of crime. (436 US 512, 56 L ed 2d 500)

MICHIGAN V CLIFFORD

In another landmark decision, *Michigan v. Clifford* (464 US 287, 78 L ed 2d 477, 104 S Ct 641), the United States Supreme Court again addressed the issue of fire investigators involved with a fire scene investigation and the elements surrounding the Fourth Amendment. As with the *Michigan v. Tyler* decision, *Michigan v. Clifford* dealt with a warrantless entry under exigent circumstances by firefighters to extinguish the fire. The firefighters may remain on scene after the fire is suppressed in case hidden fires may ignite. *Michigan v. Clifford* also stated that firefighters can also remain on the fire scene for a reasonable amount of time to determine the fire's origin and cause. However, *Michigan v. Clifford* differs from *Michigan v. Tyler* in the scope of the fire investigator's search of the fire scene after determining the origin and cause of the fire and after fire and police personnel have left the scene.

In order to understand the decision of the Supreme Court in *Michigan v. Clifford,* a look into the fire and its investigation will be studied.

The Fire and Investigation as to Origin and Cause

At approximately 5:42 AM on October 18, 1980, personnel from the Detroit Fire Department arrived at a residential structure fire. The fire was suppressed and personnel from both the police and fire departments left the scene soon after 7:00 AM on October 18th. During the fire-fighting activities, personnel found a Coleman gas can in the basement. The can was removed from the basement and placed outside by the garage door. During the firefighters' stay at the fire scene, the owner/occupant could not be found.

After completing other assignments, a Detroit fire investigator, along with his partner, arrived at the Clifford's fire damaged residence at approximately 1:00 PM, on October 18th. Upon arrival the investigators were met by workers who were securing the damaged structure with boards and pumping water out of the basement. A neighbor informed the investigators that he notified the out of town Cliffords about the fire. The Cliffords instructed the neighbor to notify the insurance agent to have the home properly secured.

Outside, near the garage door, investigators found the Coleman fuel can that was taken out of the basement by the firefighters. The can was documented and taken into evidence.

At approximately 1:30 PM, on October 18th, the investigators entered the Clifford's residence, without consent or a warrant, to conduct an investigation as to the origin and cause of the fire.

The investigation was initiated in the basement where the investigators immediately found several items of evidence to indicate arson. Evidence found included the strong odor of fuel, two additional Coleman fuel cans under the stairway, a Crockpot plugged into an electrical timer that in turn was plugged into an electrical outlet. The timer was set to energize at 3:45. The timer was found to have stopped between 4:00 and 4:30. All of the evidence found in the basement was documented and taken into custody by the investigators.

Establishing the origin and cause of the fire to be in the basement, the investigators then searched the remainder of the residence. During the search of the upstairs portion of the building, more evidence which indicated arson was observed and documented.

The Cliffords were arrested and charged with the crime of arson. During the Clifford's trial, their attorney moved to suppress all the evidence seized in the warrantless search. It was felt that the Clifford's Fourth and Fourteenth Amendment rights were violated. The trial court denied the motion on the grounds that exigent circumstances justified the search. On appeal, the Michigan Court of Appeals stated that exigent circumstances did not exist, thereby, reversing the trial court's ruling.

The United States Supreme Court Decision

The state did not challenge the exigent circumstances question with the United States Supreme Court. Instead, the state wanted to do away with the administrative warrant that was needed to determine the origin and cause of the fire. The Supreme Court denied the request, stating that they were holding firm to the decision of *Michigan v. Tyler* involving the requirements of the administrative warrant according to the Fourth Amendment. The court discussed the legality of several warrantless searches that may or may not violate the Fourth Amendment requirement when entering a fire damaged building.

The first one dealt with the "reasonable privacy expectation" of victim's personal contents within a fire-damaged building. The courts decision in *Michigan v. Clifford* went along in part with the decision of *Michigan v. Tyler*. In the opinion of Justice Powell, he stated:

> People may go on living in their homes or working in their offices after a fire. Even when that is impossible, private effects often remain on the fire damaged premises. (*Tyler*, 436 US, at 505 56 L ed 2d 486, 98 S Ct 1942) Privacy expectations will vary with the type of property, the amount of fire damage, the prior and continued use of the premises, and in some cases the owner's efforts to secure it against intruders. Some fires may be so devastating that no reasonable privacy interests remain in the ash and ruins, regardless of the owner's subjective expectations. The test essentially is an objective one...(464 US 292, 78 L ed 2d 483)

In the Clifford's fire, the court felt that the search made by the fire investigators was not a continuation of the search by the firefighters during their suppression activities. Also, during this five-hour time span, the Cliffords took steps to secure their interests. Therefore, without a warrant or consent, it was an unreasonable search by the fire investigators, violating the Clifford's Fourth Amendment rights. As Justice Powell stated:

> ...the interim efforts to secure the burned out premises and the heightened privacy interest in the home distinguish this case from Tyler. At least where a homeowner has made a reasonable effort to secure his fire damaged home after the blaze has been extinguished and the fire and police units have left the scene, we hold that a subsequent post fire search must be conducted pursuant to a warrant, consent, or the identification of some new exigency. (464 US 297, 78 L ed 2d 486)

The next issue addressed by the Court dealt with the exigent circumstance for firefighters entering a building without consent or warrant to extinguish a fire, and the exigent circumstance to determine the origin and cause. The decision in *Michigan v. Clifford* affirmed the decision from *Michigan v. Tyler*. The Court stated:

> ...in Tyler we held that once in the building, officials need no warrant to remain for a "reasonable time" to investigate the cause of the blaze after it has been extinguished. (464 US 293 78 L ed 2d 483)

As in *Michigan v. Tyler,* the Court stated that once the reasonable amount of time has elapsed and the police and fire personnel left the scene then a warrant or consent is needed to gain entry back onto the scene.

Michigan v. Clifford, as well as *Michigan v. Tyler*, stated that if a warrant is required in a fire scene search, then the object of the search determines what type of warrant will be issued. If the fire investigator needs a warrant to determine the origin and cause of the fire, then an administrative warrant is all that is needed. In order to serve the administrative warrant, the investigator must adhere to the requirements discussed in the United States Supreme Court case *Michigan v. Tyler*. Justice Powell stated in his opinion:

> If the primary object is to determine the cause and origin of a recent fire, an administrative warrant will suffice. To obtain such a warrant, fire officials need show only that a fire of undetermined origin has occurred on the premises, that the scope of the proposed search is reasonable and will not intrude unnecessarily on the fire victim's privacy, and that the search will be executed at a reasonable and convenient time. (464 US 294, 78 L ed 2d 484)

The Supreme Court made it very clear that the fire investigator must be aware that in the course of the investigation, if evidence of a crime is observed under exigent circumstances or an administrative warrant, then the scope of the investigation has changed from exploratory to criminal. The evidence can be seized, but the investigation must stop at this point and not continue. The evidence that was seized is now the probable cause for a criminal search warrant that will enable the fire investigator to conclude the fire scene examination and investigation. The only condition that the investigation may continue is under a consent to search. However, the Court wanted it understood that any of the two types of warrants did not give the fire investigator free reign to search the fire damaged structure. The search must be conducted within the language of the search warrant. Most importantly, the Court stated that even if under exigent circumstances evidence of criminal activity is found, then a criminal search warrant should be obtained. Justice Powell stated:

> Fire officials may not, however, rely on this evidence [criminal activity] to expand the scope of their administrative search without first making a successful showing of probable cause to an independent judicial officer.
>
> The object of the search is important even if exigent circumstances exist. Circumstances that justify a warrantless search for the cause of a fire may not justify a search to gather evidence of criminal activity once that cause has been determined. (464 US 294 78 L ed 2d 484)

Based on the search and seizure decisions made by the United States Supreme Court in *Michigan v. Clifford*, the search of the Clifford's fire was divided into two different searches. The first search involved the fire department's suppression of the fire and the finding of the Coleman gas can in the basement. The can was then placed outside by a firefighter. The Court ruled that since this can was observed in plain view by the fire fighters and by the fire investigator outside, this piece of evidence was admissible for retrial.

The second search involved the warrantless entry by the arson investigators, who collected evidence in the basement and other pieces of evidence upstairs. The Court ruled that the electric Crockpot, the timer and attached electric cord, the two Coleman gas cans and the testimony of the investigators concerning the warrantless search of the basement and the upstairs were "the product of the unconstitutional postfire search of the Clifford's residence" thereby making the search unreasonable under the Fourth and Fourteenth Amendments.

In conclusion, the Court affirmed the Michigan Court of Appeals concerning the evidence taken by the fire investigators. This evidence was seized without a warrant or consent thereby making the seizure a violation of the Fourth Amendment. The evidence thus became inadmissible for retrial. The evidence found by the fire personnel during the fire suppression activities, and seized by the fire investigators, was found in plain view. This decision made the evidence in plain view a lawful seizure, and reversed the Michigan Supreme Court's decision.

ADMINISTRATIVE WARRANT

As stated earlier, the landmark cases involved with the Administrative Warrants are *Camara v. Municipal Court of the City and County of San Francisco* and *See v. Seattle*. Prior to these cases, the case looked at by the courts concerning the administrative inspections was *Frank v. Maryland* (359 US 360, 3 L ed 2nd 877). This particular case held that the entry and inspection by a health inspector did not constitute a search within the Fourth Amendment requirements. One of the main reasons the Supreme Court decided to hear the *Camara and See* cases was as Justice White stated:

> More intensive efforts at all levels of government to contain and eliminate urban blight have led to increasing use of such inspection techniques.

Therefore, it was time to re-examine administrative inspections and decide if this type of inspection violated the Fourth Amendment. (Camara v Municipal Court of the City and County of San Francisco 387 US 525, 18 L ed 2d 933)

In the Camara case, an inspector from the Division of Housing Inspection of the San Francisco Department of Public Health went to an apartment building for an annual inspection. The inspector wanted to inspect the ground floor residence of Mr. Roland Camara. According to the occupancy permit that was issued, having a residence on the ground floor was a violation. When the inspector asked to enter the residence, Mr. Camara refused, unless the inspector had a search warrant. Two days later the inspector approached Mr. Camara for consent to search the ground floor residence. Again, Mr. Camara refused entry unless the inspector had a search warrant. Several other attempts were made and each time Mr. Camara requested to see a search warrant. Mr. Camara was then arrested and charged with "refusing to permit a lawful inspection" in violation of Housing Codes. Mr. Camara appealed to the District Court of Appeals stating that his Fourth Amendment rights were violated. The Court of Appeals stated that Mr. Camara's rights were not violated citing the case of *Frank v. Maryland.*

The United States Supreme Court, in its decision, first looked at the language of the Fourth Amendment. Justice White, who delivered the opinion of the Court, stated:

> The basic purpose of this amendment, as recognized in countless decisions of this court, is to safeguard the privacy and security of individuals against arbitrary invasion of government officials... One governing principle, justified by history and by current experience, has consistently been followed. Except in certain carefully defined classes of cases, a search of private property without proper consent is "unreasonable unless it has been authorized by a valid search warrant." (387 US 528, 18 L ed 2d 935)

The Court stressed the importance for the need of the public to be protected through the enforcement of "minimum fire, housing and sanitation standards." However, these standards could not be justified by a warrantless administrative search set forth in the Fourth Amendment. The problem in obtaining a warrant is that a probable cause must be shown. Since an inspection is a civil matter, a less restrictive probable cause was developed. Justice White stated:

...This is not to suggest that a health official need show the same kind of proof to a magistrate to obtain a warrant as one must who would search for the fruits or instrumentalities of crime. Where considerations of health and safety are involved, the facts that would justify an inference of probable cause to make an inspection are clearly different from those that would justify such an inference where a criminal investigation has been undertaken. (387 US 538, 18 L ed 2d 940)

Justice White also stated:

The warrant procedure is designed to guarantee that a decision to search private property is justified by a reasonable governmental interest. But reasonableness is still the ultimate standard. If a valid public interest justifies the intrusion contemplated, then there is probable cause to issue a suitably restricted search warrant. (387 US 539 18 L ed 2d 941)

Lastly, the administrative warrant should be used as a last resort. As the court stated, the majority of people will consent to an inspection without a warrant. Only a small percent will refuse and it is for these few that the warrant is needed.

The United States Supreme Court held that Mr. Camara's right to ask for a search warrant from the inspector was within his constitutional right under the Fourth Amendment. Mr. Camara, then, could not be convicted for refusing.

Very similar to the Camara case is the Supreme Court case *See v. Seattle* (387 US 541, 18 L ed 2d 943). The major difference between the two cases involves the type of structure being inspected. Camara's inspection involved a residential structure and See's inspection involved a commercial structure.

In *See v. Seattle*, an individual from the City of Seattle Fire Department was refused entry to inspect Mr. See's locked business warehouse. The official had no warrant or probable cause to enter the building. The inspection was part of a routine to see if businesses were complying with the fire codes. Mr. See was arrested and convicted. Mr. See claimed that the warrantless inspection would violate his Fourth Amendment rights.

In the opinion of the court delivered by Justice White, he did not make a distinction between a search involving a commercial property and a residential property. Both involve governmental intrusion that must comply to the requirements of the Fourth Amendment. Justice White stated:

> The businessman, like the occupant of a residence, has a constitutional right to go about his business free from unreasonable official entries upon his private commercial property. The businessman, too, has that right placed in jeopardy if the decision to enter and inspect for violation of regulatory laws can be made and enforced by the inspector in the field without official authority evidenced by a warrant. (387 US 543, 18 L ed 2d 946)

The Court concluded that an administrative entry into a commercial property, not open to the public, may be done by either a consent, through adjudication, or through the execution of a warrant. The reasonableness of each search as applied to the Fourth Amendment will be taken on a case by case basis. The Court also recognized that there are some regulatory techniques for licensing programs "which require inspections prior to operating a business or marketing a product."

According to the Court, Mr. See was within his constitutional right to ask for a warrant when the fire inspector wanted to gain entry into his business. The Court concluded Mr. See could not be prosecuted for exercising his rights.

The third landmark United States Supreme Court case dealing with an Administrative Inspection is *Marshall v. Barlow's Inc.* (436 US 307, 56 L ed 2D 305) This case developed when an OSHA's (Occupational Safety and Health Act of 1970) inspector requested from Mr. Barlow a search of his business, an electrical and plumbing installation business called Barlow's, Inc. The inspector stated that under Section 8(a) of the Occupational Safety and Health Act he is empowered by the Secretary of Labor to:

> ...search the work area of any employment facility within the Act's jurisdiction. The purpose of the search is to inspect for safety hazards and violations of OSHA regulations. No search warrant or other process is expressly required under the Act. (436 US 309, 56 L ed 2d 305)

Mr. Barlow inquired if the inspector had received a complaint concerning the company. According to the inspector, there was none. Mr. Barlow then asked the inspector if he had a search warrant. Again, the inspector replied, "No." At this point, Mr. Barlow refused the inspector entry, stating the warrantless entry would violate his Fourth Amendment Constitutional right. Shortly thereafter, the Secretary of Labor, Ray Marshall, petitioned a United States District Court to issue an order for Mr. Barlow to comply. After receiving the order, Mr. Barlow still refused entry and asked the same court to cease the order. A three judge court heard the case and ruled in favor of Mr. Barlow, citing the cases of *Camara and See* as major factors in the Court's decision. The Secretary of Labor appealed the decision, bringing the case up to the Supreme Court.

The first item addressed by the Court was the argument by the Secretary, that warrantless inspections to enforce OSHA regulations are reasonable within the scope of the Fourth Amendment. The Court disagreed. In the opinion of the Court, delivered by Justice White, he stated:

> This court has already held that warrantless searches are generally unreasonable, and that this rule applies to commercial premises as well as homes. These same cases [Camara and See] also held that the Fourth Amendment prohibition against unreasonable searches protects against warrantless intrusions during civil as well as criminal investigations. (436 US 312, 56 L ed 2D 311)

The Court then discussed that there are "certain carefully defined classes of cases" where warrantless searches of businesses can be reasonable. One of these cases involved certain businesses that the owner was licensed and regulated by the federal government. By accepting the license the owner accepted close government supervision. Some of the examples given in Justice White's opinion were the liquor and firearms industries. Justice White stated:

> The element that distinguishes these enterprises [liquor and firearms] from ordinary businesses is a long tradition of close government supervision, of which any person who chooses to enter a business must already be aware. A central difference between those cases [Colonnade–Liquor and Biswell–Firearms] and this one [Barlow's, Inc.] is that businessmen engaged in such federally licensed and regulated enterprises accept the burdens as well as the benefits of their trade, whereas, the petitioner [Barlow] here was not engaged in any regulated or licensed business. (436 US 313, 56 L ed 2d 312)

The Supreme Court concluded that Mr. Barlow was not involved in a regulated business and was within his constitutional right to ask for a warrant. In each of the cases mentioned, the underlining theme deals with the Fourth Amendment and the language that states, "to safeguard the privacy and security of individuals against arbitrary invasions by government officials."

To help the fire investigator write an Administrative Warrant, the following three forms can be used. The first form is the Affidavit statement of the fire investigator. This statement is a written declaration made under oath made before a magistrate. The second form describes the property being search and the purpose of the search. The third form is used to explain if evidence was taken during the search. If evidence is seized a list must be provided, describing in detail what was taken.

IN THE CIRCUIT COURT OF THE
_____JUDICIAL CIRCUIT
IN AND FOR_____COUNTY.

STATE OF_____
COUNTY OF_____

AFFIDAVIT FOR ADMINISTRATIVE WARRANT

BEFORE ME, _____(Judge)_____ , personally appeared affiant, _____(Name)_____ , a _____(Position)_____ with the _____(Agency)_____Department, who first being duly sworn makes this affidavit in support of the issuance of an administrative warrant, and, on oath, says:

1) ___State___ Statutes Section_____provides for the State Fire Marshal or his agents to conduct an investigation to determine the cause and origin of every fire in the State in which property is damaged or for which there exists probable cause to believe the fire was the result of carelessness or design.

2) That a fire occurred at:

> (Briefly describe property to be searched; include the address and time and date of the fire and comments on the occupancy and/or security of the premises; state that property has been damaged.) (Caveat: Always verify correct address. Use legal description of property rather than street address or describe location and premises with particularity.)

and the cause of the fire is presently under investigation.

3) The affiant believes that for the purpose of public safety, so as to prevent such fires from reoccurring and to provide information for training of public safety personnel, it is necessary to determine the cause and origin of the aforementioned fire.

4) Affiant believes the only effective means and methods for determining the cause and origin of the aforementioned fire is to physically enter the premises, examine the premises including the photography of same and, if necessary, remove samples of materials that may be connected to the cause and origin of the fire for further examination, analysis and/or testing.

WHEREFORE, the affiant requests this court to issue an administrative warrant for the purpose of permitting affiant and other fire and law enforcement officials, as reasonably necessary to conduct this investigation, to enter the aforementioned premises for the purpose of determining the cause and origin of the aforementioned fire. Said entry or entries will be made _____Describe when_____with whatever equipment as is necessary to make an investigation for said purpose.

_____(Affiant)_____

Sworn an subscribed to before me, this_____day of_____, 19____.

_____
County/Circuit Court Judge

IN THE CIRCUIT COURT OF THE
_____JUDICIAL CIRCUIT,
IN AND FOR _____ COUNTY.

STATE OF _____
COUNTY OF _____

ADMINISTRATIVE SEARCH WARRANT

TO ANY FIRE AND LAW ENFORCEMENT OFFICIAL OF SAID COUNTY:

THE ATTACHED AFFIDAVIT   HAVING BEEN WORN TO BY THE AFFIANT,

_____(Name)_____, _____(Department)_____

before me this day, based upon the facts stated therein, cause having been

found, in the name of the State of _____, I command that you enter the

following described place:_____

located at _____(Legal Description)_____

in _____County, State of _____, to locate the point of

origin and determine the cause of a fire that occurred therein on _____Date_____
You are authorized to enter, inspect, photograph and remove samples of material,

which may be connected to the origin or cause of the fire for further examina-

tion, analysis and/or testing. Said entry or entries will be made _____Describe_

when)___ and whatever equipment as is reasonably necessary to complete the

examination to determine cause and origin of this fire.

WITNESS my hand and seal this _____ day of _____, 19___.

_____
County/Circuit Court Judge

RETURN TO ADMINISTRATIVE WARRANT

STATE OF _____
COUNTY OF_____

 I HEREBY CERTIFY AND RETURN that by virtue of the within search warrant to me directed, I have searched for those items of physical evidence related to the origin and cause of a fire, at the place therein described.

(Use one of the following.)

1) And that I have such items of physical evidence related to the origin and cause of a fire before the Court described as follows:

(List and describe or attach separate
inventory page if needed.)

2) And that I have been unable to find such items of physical evidence related to the origin and cause of a fire.

Signed:_____
(Name)

_____
(Department)

Dated: At _____, _____, this
_____day of _____, 19___ .

ADMINISTRATIVE WARRANT INVENTORY

I, _____, in the presence of _____,

at _____

County of _____, State of _____, did seize the

following items from the above described location, on _____ 19__

_____

_____

_____

_____

_____

_____

_____

_____

_____

_____

_____

_____

_____

_____

_____

_____

_____

_____

_____

_____ Dated _____, 19__.

Signed _____

Chapter 2

CONSENT

Consent to search by definition, is the agreement by an individual either verbally or written that allows a governmental official (law enforcement, fire personnel, health inspectors, building inspectors, etc.) to search that individual's person, place or personal contents. What the individual is actually doing is giving up their constitutional rights and giving the governmental official the right to search and seize evidence without probable cause and without a warrant. This type of warrantless search is the fastest and less complicated way to gain entry after exigent circumstances are over and the reasonable amount of time has elapsed.

However, the courts will look very closely as to how the consent was given. The consent must be given freely, without duress and "not coerced by explicit or implicit means." The United States Supreme Court case, *Schneckloth v. Bustamonte* (412 US 218, 36 L ed 2d 854) is looked upon as the foundation of law for consent to search. Justice Stewart, who delivered the opinion for the majority, stated:

> The Fourth and Fourteenth Amendment require that consent not be coerced, by explicit or implicit means, by implied threat or covert force. For, no matter how subtly the coercion was applied, the resulting "consent" would be no more than a pretext for the unjustified police intrusion against which the Fourth Amendment is directed. (412 US 228, 36 L ed 2d 854)

It is up to the prosecution to prove that the evidence obtained under the "consent to search" was given freely and voluntarily. The testimony of the governmental official who obtained the "consent to search" will reveal how the official behaved him or herself in obtaining the consent and the ensuing search, and if it was done lawfully.

The courts do not have a set standard to determine if and when the consent is unconstitutional. Each case will be examined, concerning the facts and circumstances surrounding the consent to determine if it was made voluntarily or involuntarily. In the Supreme Court case *Bumper v. North Carolina* (391 US 543, 548, 20 L ed 2d 797, 88 S Ct 1788), the Court stated:

> When a prosecutor seeks to rely upon consent to justify the lawfulness of a search, he has the burden of proving that the consent was, in fact, freely and voluntarily given.

To determine the definition of "voluntarily," the Court turned to the many cases involving the "voluntariness" of a defendant's confession. The Court's conclusion of the voluntary confession was that in each particular case the Court would look at "the totality of all the surrounding circumstances, both the characteristics of the accused and the details of the interrogation" *Schneckloth v. Bustamonte* 412 US 226, 36 L ed 865). The courts therefore, applied the same principles for a "voluntary confession" to the "consent to search."

Another issue that was discussed in *Schneckloth v. Bustamonte* involves whether an individual must have the knowledge to refuse consent. The Court felt that the right to refuse consent did not have to be voluntarily given to the individual by the governmental official. The Court would, however, look at the conditions of the consent to see if voluntariness was a factor.

The mental and/or physical conditions of the individual are some of the other factors to consider when looking for the voluntary consent. A person intoxicated on alcohol or drugs may not be capable of giving consent. An individual who is uneducated or mentally retarded may not be able to give consent. Due to the fact that the official must testify as to the voluntariness of the consent, the official should examine closely the individual's physical and emotional state before executing the consent to search.

Consent may be given verbally or written. Either one will relinquish an individual's constitutional right. However, since the official must prove the consent is lawful, it is better to exhibit a written document to the court that has been signed by the individual and witnessed. With the verbal consent, the proof is in the word of the official against the individual.

WHO HAS THE RIGHT TO GIVE CONSENT?

An individual who has authority, use or control over the premises may give consent. A third party consent may be given if they have mutual use, joint occupation. The evidence seized in a third-party consent against the nonconsenting individual has been ruled constitutional. The United States Supreme Court case, *United States v. Matlock*, (415 US 164, 39 L ed 2d 242) involves third party consents. The Court stated:

> When the prosecution seeks to justify a warrantless search by proof of voluntary consent, it is not limited to proof that consent was given by the defendant, but may show that permission to search was obtained from a third party who possessed common authority over or other sufficient relationship to the premises or effects sought to be inspected. (415 US 171, 39 L ed 2d 249)

To define "common authority" the Court stated:

> Common authority is, of course, not to be implied from the mere interest a third party has in the property. The authority which justifies the third party consent does not rest upon the law of property, with its attendant historical and legal refinements, but rests rather on mutual use of the property by persons generally having joint access or control for most purposes, so that it is reasonable to recognize that any of the co-inhabitants has the right to permit the inspection in his own right and that the others have assumed the risk that one of their number might permit the common area to be searched. (415 US 171, 39 L ed 2d 250)

Co-inhabitants are those individuals who share a common interest of the premises. They can be roommates, spouses or live-in lovers. The courts are going to look closely at the third party consent to see if it meets the "common authority" test. Do the co-inhabitants share in the rent, how long have they been together, does each occupant possess a key, is there a contract between the parties, are personal contents left on the premises? These are some of the questions the courts will ask.

Just because there are parents on the premises does not give the governmental official the right to enter a child's room with the parent's consent. If the child is paying rent, keeps the door to his or her room locked, and holds the only key, or does anything to protect his or her expectations of privacy, then the official must respect the child's Fourth Amendment rights and abide by the requirements set forth. If the parents enter the room of the child routinely, to clean or visit, and there is a common interest shared by both the parent and the child, then the parent may give consent.

On the other hand, a minor generally cannot give consent to officials to enter the parent's home. The child does not meet the "common authority" test; the child has use of the area, but does not have the authority.

The question of whether a guest or visitor may give consent depends on several factors. If the guest's stay is of short duration, then they may not give consent at the host's premises. However, the host, because he or she has use and control, may consent to a search involving the guest or visitor. The guest who has a long stay and has developed an area of privacy may not have that area searched on the consent of the host.

When the landlord or owner rents out an apartment, they give up the right of use to the tenant. The tenant, in turn, has the expectation of privacy. Therefore, based on the "common authority," the landlord or owner does not have the right to give consent to search the occupied leased unit. If the tenant abandons the leased unit for good, then the unit reverts back to the landlord and/or owner who then may give consent. Also, the landlord and/or owner may give consent to common areas of apartment buildings such as hallways, recreation rooms, laundry rooms, or any other areas that are under the control of the landlord and/or owner. Be very careful when the landlord and/or owner states that they can enter the unit under lease for inspection. Many of the lease agreements have this written in the contract. The tenant still has an expectation of privacy and, therefore, one of the requirements of the Fourth Amendment concerning search and seizure must be met.

A guest who acquires a room from a hotel or motel does so with the expectation of privacy for that area. Even with the knowledge that maids or work personnel may enter, a governmental official cannot get consent from the hotel/motel management. The United States Supreme Court, in the case of *Stoner v. California* (376 US 489, 11 L ed 2d), stated:

When a person engages a hotel room he undoubtedly gives "implied or express permission" to "such persons as maids, janitors or repairmen" to enter his room in the performance of their duties.

A guest in a hotel room is entitled to constitutional protection against reasonable searches and seizures. That protection would disappear if it were left to depend upon the unfettered discretion of an employee of the hotel. (376 US 489, 11 L ed 2d 361)

An individual employer may give governmental officials consent to search the premises used by employees. That includes employee's lockers that the employer has a master key for entry and with the employee's understanding that the employer may enter at any time. If, on the other hand, each locker is independently secured by the employee, then that employee's right to privacy is expected.

An employee who does not have authority on the job cannot give consent to search the employer's premises. This type of employee is the average worker or a person temporarily put in charge. The employee who is a manager or an individual who has authority and is left in charge may have the authority to give the consent.

It is felt by the courts that if an individual gives consent, that individual may at any time during the search ask to have it withdrawn. Any evidence found after the statement to withdrawal will be inadmissible in court. The courts have a strong opinion concerning the nullification of consent, especially since the Miranda case where the defendant may refuse to answer any questions. If consent was given by the "third party" and the principle occupant or owner of the property arrives, then the governmental official must ask that individual for consent before continuing.

Whenever consent is given, it is not a consent for future searches. If the governmental official wants to get back onto the scene, the official must get another consent to search. The official can state, and have it in writing on the consent form, that the consent will be ongoing for as long as it is needed. This also can be withdrawn at anytime.

The governmental official cannot expand the area of the search other than what was asked of the individual giving consent. Entering an individual's residence for an interview is not an open invitation to search the complete residence. If the governmental official has the consent to look around the premises this does not give the official an open invitation to open and search closets, dresser drawers, or closed containers without another consent. It is best to remember that when asking for consent, the official should keep the search within the area specified and not make it an exploratory search. This also applies to a consent to search of an individual's person.

CONSENT USING ACCELERANT DETECTION CANINES

As part of an investigation, accelerant detection canines can be used as a tool to search fire-damaged property for the presence of accelerants. The concept of the accelerant detection canine is based on the canine's sensitive sense of smell. The canine is trained by the handler to detect the presences of flammable/combustible accelerants. This concept is very similar to those used by law enforcement canine handlers to detect narcotics. In fact, so similar are the two types of canine searches, it is believed that the existing case laws for the narcotics detection canines will also apply to the accelerant detection canines. With this thought in mind, there has been a recent state appellate court decision in the state of Florida that involves a narcotics detection dog during a search without consent or a search warrant. This decision could have an impact on how K-9 accelerant dogs can enter and search a fire scene. To understand how an accelerant detection canine search may become a Fourth Amendment violation, let's examine the Florida case.

In *Artuor Dominguez v. The State of Florida*, Dominguez was arrested and convicted for trafficking in cocaine. The evidence was discovered and seized during a search of Dominguez's apartment for narcotics. After giving the police officers consent, a narcotics detection dog was requested and brought to the apartment. During the search the dog alerted to an area behind a sink where narcotics were found inside a wall.

On appeal, the Third District Court of Appeals felt that the search of the apartment using the narcotics dog exceeded the scope of the search for which consent had been given. The opinion of the court written by Judge Cope stated:

> The present search was the search of a home. "A private home, as here is an area where a person enjoys the highest reasonable expectation of privacy under the Fourth Amendment... and... accordingly, the factors bearing on the voluntariness of a consent to search a home must be scrutinized with special

care" *Gonzales v. State*, 578 So. 2d 729, 734 (Fla. 3d DCA 1991) (citation omitted). Here, the standard of objective reasonableness was not met. The defendant consented to an entry and search by the officers. However, the officers were not accompanied by the drug detection dog when they obtained the search. The officers did not make a specific request for consent to the use of a drug detection dog. In our view, the consent given to the officers did not by implication include the subsequent entry by the dog and its handler. In the absence of an express consent or circumstances from which consent to the use of the drug detection dog could be reasonably implied, the officers exceeded the scope of the consent given.

Applying this decision to a fire scene search by an accelerant detection canine, the fire investigator must consider several Fourth Amendment requirements established in the United States Supreme Court case *Michigan v. Clifford.* First, in almost all cases the canine will be requested and arrive at the scene after the "exigent circumstance is over." Therefore, any other searches conducted after the exigent circumstance must be conducted either through a search warrant or with a consent to search. The majority of the people who would give a written consent to search a fire scene would also give consent to use the accelerant detection canine. To avoid any grey area the language of the written consent should also contain a paragraph that informs the individual that he or she is authorizing the consent to search and use of the accelerant detection canine during the fire investigation. If a search warrant is obtained, make sure that part of the language of the PC (Probable Cause) affidavit contains the use of the accelerant detection canine and the purpose for such a search.

The following Consent to Search forms can be photocopied and written out at the fire scene. Figure 2-1 is the most up-to-date Consent to Search form that can be used. This form also includes the consent that authorizes the use of an accelerant detection canine. Figures 2-2 and 2-3 are various ways a Consent to Search form can be written.

CONSENT TO SEARCH AND REMOVE EVIDENCE:

I, _____ , the _____
    (Person giving consent)   (Owner, tenant, manager, etc.)

of the _____ located at _____
    (Residence, business, vehicle   (Address)
    vessel, etc.)

_____ do hereby freely

and voluntarily give my consent to _____
                        (Name of official)

of the _____ and any other fire
    (Agency)

Official, investigator or law enforcement officer participating in the investigation of this fire incident, to enter and search the property described above and the surrounding areas of the premises, including any other structures or vehicles situated on or adjacent to the property, to examine and remove evidence relating to the fire which occurred on or about _____
                                         (Date and time)

I specifically give my consent and authorize these persons to inspect and remove any items of evidence which may be related, directly or indirectly, to the investigation of the circumstances and cause of the fire, and to submit the evidence to examination, analysis, and/or testing. I specifically give my consent and authorize these persons to use an accelerant detection canine (fire dog) to assist and participate in the search of the premises, surrounding areas, other structures and vehicles. This consent shall remain in effect and shall authorize subsequent entry and removal of evidence as often as may be necessary to complete the investigation of this fire incident or until the consent is revoked in writing.

Signature _____

Current address _____

_____

Phone _____

Date _____   Witnessed by _____

Time _____

RE: Agency File No. _____

CONSENT TO INVESTIGATE FOR FIRE CAUSE

I, _____ , the
  (Name of person giving consent)

_____ of a _____
  (Owner, tenant, etc.)   (Type of building, vehicle or vessel)

located at _____
         (Complete address - including city or township)

_____ do hereby freely and voluntarily

give my consent to _____ of the _____
                  (Name of official)   (Department)

or his designate to examine the above described premises for the purpose of
discovering and determining the cause of a fire which occurred at the above
described premises on _____. I hereby further give my consent
to the aforementioned persons to remove any and all items, which may be related to the cause of the aforementioned fire for further examination,
analysis and/or testing.

Signed _____

Witness _____

Date _____

Time _____

CONSENT TO INVESTIGATE FOR FIRE ORIGIN AND CAUSE

BEFORE ANY SEARCH IS MADE, YOU MUST UNDERSTAND YOUR RIGHTS:

1. You may refuse to consent to a search and may demand that a search warrant be obtained prior to any search of the premises described below.

2. If you consent to a search, anything of evidentiary value seized in the course of the search can and will be introduced into evidence in court against you.

3. You have the right to decline this consent at any time.

I HAVE READ THE ABOVE STATEMENT OF MY RIGHTS AND AM FULLY AWARE OF MY RIGHTS.

I HEREBY CONSENT TO A SEARCH WITHOUT WARRANT BY _____

AND/OR _____ OFFICERS OF THE _____

OF THE FOLLOWING: (Describe premises or automobile.)

_____

_____

_____

I HEREBY AUTHORIZE THE SAID OFFICERS TO SEIZE ANY ARTICLE WHICH THEY MAY DEEM

TO BE OF EVIDENTIARY VALUE.

THIS STATEMENT IS SIGNED OF MY OWN FREE WILL WITHOUT ANY THREATS OR PROMISES

HAVING BEEN MADE TO ME.

_____   _____
(Witness)   (Signature of subject)

_____   _____
(Witness)   (Date and time)

To sum up, governmental officials who are requesting a consent to search must remember that there are three important items for a valid consent.

1. The consent must be voluntary.
2. The person giving the consent has the use and control of the place to be searched.
3. The search is within the scope of the consent obtained.

If there is any question as to the validity of the consent to search of the fire scene, halt the investigation and either get an administrative warrant or a criminal search warrant.

Chapter 3

THE PLAIN VIEW DOCTRINE

The Plain View Doctrine involves objects of evidence that are observed in the open, with an unobstructed view by governmental officials. Where searches are a government's action "that intrudes into a person's reasonable expectation of privacy," they are in violation of the Fourth Amendment. Looking directly at evidence, unconcealed, is not considered an intrusion into the expectation of privacy. Therefore, plain view does not constitute a search according to the Fourth Amendment. The object of evidence observed in open view can be seized without a warrant and used in a court of law, as long as the seizure is within certain guidelines set forth in the Fourth Amendment and the decisions formed by the United States Supreme Court.

As the cases became more intricate involving the Plain View Doctrine, the Supreme Court started putting limitations as to when a governmental official could use the Plain View Doctrine.

One of the first cases in which the United States Supreme Court recognized evidence to be in "plain view" was in *United States v. Lee* (274 US 559, 71 L ed 1202). The opinion of the court, delivered by Justice Brandeis, did not use the exact wording "plain view." The evidence was eluded to when a United States Coast Guard official stated that he observed contraband on the deck of a boat when his searchlight illuminated the evidence. Because the contraband was observed on the open deck, a "no search" was declared by the court. This made the search of the contraband admissible in court.

Two other Supreme Court cases followed involving evidence observed in the open. The first case, *United States v. Marron* (275 US 192, 72 L ed 231), dealt with evidence seized not described in a search warrant. The evidence taken, a ledger and some bills, were in "plain view" when an individual was arrested for openly committing a crime in the presence of the officers executing the search warrant. The second case, *Go-Bart Importing Co. v. United States* (282 US 344, 75 L ed 374), dealt with evidence that was seized by Federal agents during a lawful arrest by warrant and believed upon a probable cause that the individuals arrested were engaged in the commission of a crime.

The phrase "plain view" was first used by the Supreme Court in 1932 in the case, *United States v. Lefkowitz* (285 US 452, 76 L ed 877). This case involved a warrantless search of a business by law enforcement officials who through a lawful arrest, searched and seized the contents of a closed and unlocked cabinet and desk along with the discarded paper in the wastepaper baskets. The defendants were convicted in part from the evidence taken from the cabinet, desk and wastepaper basket. The Court felt that this was an exploratory search of the complete business. In the opinion of the Court, Justice Butler stated:

> The searches were exploratory and general and made solely to find evidence of respondent's guilt of the alleged conspiracy or some other crime. (285 US 465, 76 L ed 883)

The Court also felt that a search contemporaneously with an arrest cannot be greater than a normal criminal search warrant. (In the criminal search warrant, probable cause and a description of the premises and property to be taken must be presented to a neutral party.) The Court also made a distinct statement concerning evidence in "plain view" from the court case *Marron v. United States* (275 US 192, 72 L ed 231) and the evidence taken in this case. Justice Butler stated:

> These searches and seizures are to be distinguished from the seizure of a ledger and some bills that was sustained in the Marron case...The ledger and bills being in plain view were picked up by the officers as an incident of the arrest. No search for them was made. (285 US 465, 76 L ed 882)

The court decided then, that the evidence observed in plain view was distinguished from those that were hidden from view and had to be located through a search. The evidence in plain view was admissible, any other evidence found was inadmissible. The court went on to say that the search and seizure could not be justified as incident of an arrest due to the broad area covered by the arresting officers.

This case is important to fire investigators. The courts look at *United States v. Lefkowitz* when items of evidence have been taken from secured, private places, without a valid consent, the lack of exigent circumstances and when it is outside the probable cause affidavit. Seizures under these circumstances are, according to the court, an invasion into a person's right of personal security and liberty that was given through the Fourth Amendment. It protects all those unlawful as well as those that are innocent. If the investigator knows of evidence that may be contained within a closed cabinet or of paperwork that may help the investigator determine if the crime of arson has been committed, then it must be stated in the search warrant or in consent.

The first case that the United States Supreme Court specified constitutional requirements for "plain view" was in the case *Harris v United States* in 1968 (390 US 234 19 L ed 2d 2067). In part, the opinion of the Court stated:

> ...It has long been settled that objects falling in the plain view of an officer who has a right to be in the position to have that view are subject to seizure and may be introduced in evidence. (390 US 236, 19 L ed 2d 1069)

In 1971, The United States Supreme Court case, *Coolidge v. New Hampshire* (403 US 443, 29 L ed 2d 564). set the precedent for decisions concerning the Plain View Doctrine. The State of New Hampshire tried to justify the warrantless seizure of a murderer's automobile in three different ways. The first theory involved the search and seizure of evidence incident to a valid arrest. The second theory involved the Carroll Doctrine (a warrantless search of an automobile that is lawful if the officer has probable cause to believe that items of a criminal activity are within the vehicle. The items may be seized based on the fact the vehicle could be moved before a warrant is issued). The third theory also was in support of a warrantless search and seizure of the automobile based on the fact that it was used in the criminal activity and was in plain view of law enforcement officers.

The court felt that the problem with the third theory was the lack of clarity in identifying the prerequisites for the Plain View Doctrine. In reviewing all the past Supreme Court cases involving "plain view," the court found a common factor. In the opinion delivered by Justice Stewart, he stated:

What the "plain view" cases have in common is that the police officer in each of them had a prior justification for an intrusion in the course of which he came inadvertently across a piece of evidence incriminating the accused. The doctrine serves to supplement the prior justification whether it be a warrant for another object, hot pursuit, search incident to lawful arrest, or some other legitimate reason for being present unconnected with a search directed against the accused and permits the warrantless seizure. Of course, the extension of the original justification is legitimate only where it is immediately apparent to the police that they have evidence before them; the "plain view" doctrine may not be used to extend a general exploratory search from one object to another until something incriminating at last emerges. (403 US 466, 29 L ed 2d 583)

The court then placed three constitutional limits on governmental officials who use "plain view" in a warrantless seizure. First, as in *United States v. Harris*, the law enforcement officer must make the intrusion lawfully and be in the position to observe the particular item. Second, the discovery of evidence in plain view must be inadvertent. And third, it must be immediately apparent to the law enforcement officers that the item before them is evidence.

The first limitation imposed by the plurality of the Supreme Court in *Coolidge v. New Hampshire* stated that "plain view" by itself does not justify a warrantless seizure. It is important to note that the governmental official must be in the area lawfully and see the evidence inadvertently. If the governmental official has probable cause to believe that items of evidence are in "plain view" before actually seeing the evidence, then a warrant must be obtained before a search is executed.

The second limitation in *Coolidge v. New Hampshire* stated that a law enforcement officer must find the evidence without knowing in advance the location of the evidence. Also, law enforcement officers cannot plan a warrantless search and seizure by having themselves in a better position to observe the evidence. As stated in limitation number one, if probable cause exists, then a warrant must satisfy the search and seizure to become lawful.

The third limitation in *Coolidge v. New Hampshire* states that when governmental officials observe items in plain view, they must immediately know that those items are evidence of a crime or contraband.

In the *Coolidge v. New Hampshire* case, the Court could not justify the "plain view" doctrine in seizing the defendant's automobile since they, the police, had plenty of time to obtain a warrant. Also, the police intended to seize the automobile, thereby defeating the limitations concerning the discovery of the evidence being inadvertent.

The next major Supreme Court decision involving the "plain view" doctrine was *Texas v. Brown* (460 US 730, 75 L ed 2d 502) in 1983. This case involved an individual who was arrested at a routine driver's license checkpoint. The arresting law enforcement officer observed the driver, and only occupant, withdraw his hand from a pocket and drop an "opaque, green party balloon." The balloon with a knot about one-half inch from the top, fell onto the front seat next to the driver. When asked to produce a driver's license, the driver opened the vehicle's glove compartment where the officer observed more balloons in an open bag along with several plastic vials containing a white powder. The officer then asked the driver to step out of the vehicle. The driver did so and at that point the officer reached in the vehicle and seized the green balloon on the front seat. The officer, believing the white powder to be narcotic, arrested the driver.

The defendant wanted the evidence suppressed at his trial based on the fact that the seizure of evidence did not meet the limitations of the "plain view doctrine." Of special concern to the defendant was the limitation concerning "immediately apparent" that was established in *Coolidge v. New Hampshire.*

The first topic discussed by the Court in a plurality opinion encompassed the limitations set forth in *Coolidge v. New Hampshire.* These limitations would be a "point of reference" for further discussion on the issue.

Another item the Court wanted to clarify was the language of the limitation "immediately apparent" in *Coolidge v. New Hampshire.* Since the decision in *Coolidge v. New Hampshire*, many cases appeared before the Court involving the language of the limitation "immediately apparent." The Court felt that the intent was too ambiguous and rationalized that this was a good time to clarify the phrase. In a plurality opinion delivered by Justice Rehnquist he stated:

> The use of the phrase, "immediately apparent" was very likely an unhappy choice of words, since it can be taken to imply that an unduly high degree of certainty as to the incriminatory character of evidence is necessary for an application of the "plain view" doctrine. (*Texas v. Brown* 460 US 741, 75 L ed 2d 513)

To define "immediately apparent," Justice Rehnquist restated from several other Supreme Court cases. The following is one of those cases:

> The seizure of property in plain view involves no invasion of privacy and is presumptively reasonable, assuming that there is no probable cause to associate the property with criminal activity. (*Texas v. Brown*, 460 US 41, 75 L ed 2d 513)

Following this statement, Justice Rehnquist reaffirmed the Court's belief on probable cause.

> "Probable cause is a flexible, common sense standard. It merely requires that the facts available to the officer would "warrant a man of reasonable caution in the belief," (site omitted) that certain items may be contraband or stolen property or useful as evidence in a crime; it does not demand any showing that such a belief be correct or more likely true than false. (*Texas v. Brown*, 460 US 742, 75 L ed 2d 514)

The Supreme Court, in deciding the *Texas v. Brown* case, put the evidence seized under the "Plain View Doctrine" to the limitation test of *Coolidge v. New Hampshire*:

The first limitation passed because the arresting officer was in a lawful position, checking driver's licenses at the checkpoint.

The second limitation passed when the arresting officer inadvertently observed the balloon on the front seat while checking for a driver's license.

The third limitation also passed because the officer had probable cause to believe the balloon contained an unlawful substance. This was established through the officer's testimony that he had observed similar types of evidence during the participation of past narcotics arrests and through conversation with fellow police officers.

The Supreme Court then, in its decision, upheld the seizure and arrest as lawful.

The following year, in 1984, the United States Supreme Court case, *Michigan v. Clifford,* made an exception to the "Plain View Doctrine" for fire investigators. The Court felt that certain items of evidence may be hidden from view under the fire debris. This evidence may come into view during the investigation as to the origin and cause of the fire. Justice Powell stated in his opinion of the Court:

The plain view doctrine must be applied in light of the special circumstances that frequently accompanies fire damage. In searching solely to ascertain the cause, firemen customarily must remove rubble or search other areas where the cause of fires are likely to be found. An object that comes into view during such a search may be preserved without a warrant. (*Michigan v. Clifford* 464 US 295, 78 L ed 2d 485)

The last Supreme Court case to be discussed in the "Plain View Doctrine" is *Arizona v Hicks* (480 US 321, 94 L ed 2d 347, 107 S ct 1149) in 1987.

This case involved a shooting incident where a bullet went through an apartment floor, injuring the occupant of the apartment directly below. Police entered the apartment where the shooting occurred to look for the shooter, other shooting victims and for weapons. During the search, three weapons were recovered along with some expensive pieces of stereo equipment. One of the police officers felt that the equipment looked out of place. Suspecting the stereo equipment might be stolen, one of the police officers moved some of the equipment to obtain serial numbers. Via a telephone at the apartment, the officer ascertained the equipment was recently taken in an armed robbery and was immediately seized. Other stereo equipment was also believed to be taken from the same robbery and a warrant was issued for the seizure. The occupant was then charged for the armed robbery based on the evidence (stereo equipment) found at the apartment.

The question faced by the Supreme Court in this case was whether less than probable cause is enough to seize evidence under the "Plain View Doctrine." In the opinion of the Court, delivered by Justice Scalia, he stated that the copying of the serial numbers of the stereo equipment did not constitute a seizure due to the fact that it did not interfere with the "respondent's possessory interest in either the serial numbers or the equipment." However, the officer moving the stereo equipment was a separate search from that of the initial lawful search (looking for the shooter, victims and weapons). The areas of the stereo equipment that were in plain view did not constitute an invasion of privacy. The invasion occurred when the officer moved the equipment to find the serial numbers that were hidden from view. Justice Scalia stated:

Merely inspecting those parts of the turntable that came into view during the latter search would not have constituted an independent search, because it would have produced no additional invasion of respondent's privacy interest. But taking action, unrelated to the objectives of the authorized intrusion, which exposed to view concealed portions of the apartment or its contents, did produce a new invasion of respondent's privacy unjustified by the exigent circumstance that validated the entry. It matters not that the search uncovered nothing of any great personal value to the respondent–serial numbers rather than (what might conceivably have been hidden behind or under the equipment) letters or photographs. A search is a search, even if it happens to disclose nothing but the bottom of a turntable. (Arizona v. Hicks, 480 US 325, 94 L ed 2d 354)

The last question to be answered by the Court was whether or not the search for the serial numbers was reasonable. To help answer this question, the Court placed another limitation on the "Plain View Doctrine." The Court stated through Justice Scalia's opinion:

> We have not ruled on the question whether probable cause is required in order to invoke the "Plain View" Doctrine. *Payton v. New York* (citation omitted) suggested that the standard of probable cause must be met, but our later opinions in *Texas v. Brown* (citation omitted) explicitly regarded the issue as unresolved.
> We now hold that probable cause is required. To say otherwise would be to cut the "Plain View Doctrine" loose from its theoretical and practical moorings. (*Arizona v. Hicks* 480 US 326, 94 L ed 2d 355)

With this limitation now imposed, the Court concluded probable cause was required that the stereo equipment was stolen for a lawful search and seizure. Even the officer stated in testimony he had only a "reasonable suspicion." With the limitation of probable cause nonexisting, the third limitation of "immediately apparent" was not satisfied.

What this did to governmental officials involved in the "Plain View Doctrine" was to have that official believe probable cause existed on an item that might be evidence of a crime or contraband. Some other form of intrusion cannot be made to justify the seizure. For example, during a fire scene investigation, a gun is discovered. The moving of the gun to get a serial number to see if the gun is stolen or used in a crime is not justified. Without probable cause to believe the gun was used in a criminal activity, the seizure cannot be used in a court of law. This can also apply to notebooks, letters, files, videotapes, etc.

The things to remember in order to have a lawful search and seizure under the "Plain View Doctrine" are:

1. You must be in an area lawfully, either by a warrant or one of the exceptions to the warrant.

2. You must "inadvertently" come across the evidence. No other prior knowledge must exist that the evidence was there.
3. The evidence must be "immediately apparent" that it is of evidentiary value.
4. You must have probable cause to believe the item of evidence was part of a crime or contraband.

Chapter 4

ABANDONMENT

Another exception to the Fourth Amendment requirement encompassing search and seizure is called abandonment. In general, abandonment deals with the relationship between property and its owner, when the owner voluntarily gives up all rights, title and possession, with the intent of discarding the property, not giving the property to anyone and not reclaiming it. With the owner giving up the right of ownership, the right of expectation of privacy afforded by the Fourth Amendment is also relinquished. With the expectation of privacy discarded, the search and seizure of evidence in a nonconstitutional protected area is lawful.

There are two types of property that can be abandoned. The first type involves the discarding of property that consists of moveable property that can be transported by the owner. Such property can be personal contents, objects that can be moved from place to place, and any type of mobile transportation. The second type involves the discarding of property that is nonmoveable. Nonmoveable property would consist of a structure, either in part (such as a motel room, a rented office or apartment) or as a whole (such as a dwelling or commercial building).

One of the first major United States Supreme Court cases concerning abandonment was in *Hester v. United States* (265 US 57, 68 L ed 898). The case involved the seizure of evidence by revenue officers who, acting on obtained information concerning illegal whiskey, went to the home belonging to Hester's father. When the revenue officers approached the house, from a concealed spot, they observed a vehicle drive to the dwelling. Hester came out carrying a glass container and handed it to the occupant of the vehicle. An alarm was given and both men carrying glass containers ran across an open field. One of the revenue officers gave chase and fired his weapon. Hester dropped his glass container in the open field. The container broke, but enough of the container and its liquid contents were available for seizure into evidence. The other revenue officer went to the house and found a broken glass container holding a liquid outside. The officer seized as evidence both the broken container and its liquid. Arrests were made by the officers without a search or arrest warrant. Hester's defense was based on the fact that the search and seizure of the evidence violated his constitutional right under the Fourth Amendment concerning "unreasonable searches and seizures." In the opinion of the Supreme Court, delivered by Justice Holmes, the seizure of evidence did not violate the Fourth Amendment rights of Mr. Hester because he abandoned those rights when he discarded the glass container. Justice Holmes stated:

> It is obvious that, even if there had been a trespass, the above testimony [the arresting officer stating they had no warrant to search, to seize or to make an arrest] was not obtained by an illegal search or seizure. The defendant's own acts, and those of his associates, disclosed the jug, the jar and the bottle; and there was no seizure in the sense of the law when the officers examined the contents of each after it had been abandoned. (265 US 58, 68 L 900)

Throughout the years the courts have made decisions concerning the discarding of moveable personal property and objects, objects thrown out of moving vehicles, items discarded while on foot, to the search of curbside garbage. The key to all cases involving abandonment is whether or not the presence of the governmental official is lawful. The product of a misconduct by governmental officials will in most cases make any evidence found, inadmissible in court.

Another key element of abandonment centers on the idea did the owner have the intent of voluntarily discarding the property. This was the issue in a recent Connecticut Supreme Court Decision involving the taking into evidence of a victim's burnt clothing and having it analyzed for acelerants. The events in this case, *State of Connecticut vs. Joyce* (Conn. Sup Ct, No. 14708, 3/16/94). are the following. Firefighters and paramedics responded to a call involving an explosion and house fire. Upon arrival, fire units found the house fully involved and Mr. Wallace Joyce standing in a river near the fire scene with severe, lifethreatening, thermal injuries. Initial treatment of Mr. Joyce resulted in the paramedics cutting off Mr. Joyce's clothing to expose his thermal injuries. The clothing was left on the ground and Mr. Joyce was transported to the hospital. A detective on the fire scene picked up the clothing and took it into custody. This was a standard procedure used by the police department to safeguard and return to the proper owner personal property that was found. The detective stated at trial that during the recovery of the clothing Mr. Joyce was not a suspect. After the fire was determined to be an arson, Mr. Joyce became a suspect, at which time the detective gave the clothing to a fire marshal who in turn submitted the clothing for "chemical testing." Laboratory results of the clothing indicated the presence of gasoline. The laboratory results, along with other evidence, convicted Mr. Joyce of arson.

On appeal, Mr. Joyce felt that his Fourth Amendment rights and other rights found in the Connecticut Constitution were violated when the clothing was analyzed without consent or a search warrant. The opinion of the Court given by Justice Berdon stated that there were three issues that must be determined in this case:

> We must determine (1) whether there was a reasonable expectation of privacy in the clothing, (2) whether the testing of the clothing at the state laboratory constituted a search, and (3) if so, whether the circumstances of this case fall within a recognized exception to the warrant requirement.

The majority of the court felt that the defendant *did not* abandon his clothing under the circumstances and therefore still retained his right to privacy. Justice Berdon stated:

> We conclude that the defendant adequately exhibited his subjective expectation of privacy, as he "merely left his property behind him, more or less of necessity, making no attempt, however, to discard it or disassociate it from himself." (*State v. Philbrick*, 436 A 2d 844, 855 (Maine SupJudCt 1981)

Second, the Court made it quite clear that the police action of seizing the clothing was accepted in that they were protecting the rights of the defendant, Mr. Joyce. However, when the clothing was chemically analyzed, the court felt that this was "capable of determining a multitude of private facts about an individual" and therefore constituted a search.

Since the laboratory analysis was a search, it had to meet the requirements for a warrantless search. The state could not give a warrantless exception to the search and wanted the Court to consider the search to that of search that is conducted when a vehicle is at a police station, in custody. The Court did not agree and felt that the rights of every citizen, including criminals, to privacy, must be protected. Justice Berdon stated:

> Our concern for the right to personal privacy and our reference for the warrant to protect that privacy was recently underscored when we pointed out that even if the police act without a warrant under the emergency exception, once that emergency ceased to exist, the police must terminate their intrusive conduct.

The Court concluded that the testing of the clothing for accelerants without a consent or a search warrant was an unreasonable search.

In cases involving vehicles of transportation, the courts have made several decisions of abandonment. In 1980, in a South Carolina court case, *South Carolina v. Lemacks* (268 SE 2d 285) it was decided:

> Leaving the vehicle unattended, with keys readily available, and parked in such a manner as to constitute hazard to vehicular traffic constitutes abandonment for constitutional purposes and removes it from protection of Fourth Amendment. (*Searches and Seizures*, 68 Am Jur 2d supp. 182)

In another case from Texas in 1982, *Hudson v. Texas* (642 SW 2d 562) the court stated:

> ...the defendant voluntarily abandoned his car and therefore lost standing to complain of an illegal search and seizure, not withstanding that he was shot while fleeing from the car, where he left the car with police by the side of the highway at 3:00 AM and made no actual effort to reclaim it; once abandoned, the police could lawfully impound the car and evidence located in it was properly admitted at trial." (Searches and Seizures, 68 Am Jur 2d Supp. 182)

As fire investigators, discarded objects, such as containers of accelerates, charred clothing or any other piece of evidence to indicate the crime of arson, may be lawfully seized and tested (searched) if they meet the requirements for abandonment or fall under any of the other requirement for a warrantless search. (If the fire investigator believes it to be in a grey area or cannot meet the warrantless search requirements, then a search warrant or consent should be obtained.) Fire scene investigation of vehicles may also use abandonment as an expectation to a warrantless search and seizure, or any of the other exceptions to the warrant requirement if applicable.

Under the second type of abandonment, nonmoveable structures can be abandoned either in part or as a whole. When a hotel/motel room, apartment, a room in a rooming house or an area is leased, the abandonment occurs when personal contents are removed, the bill is paid, and the occupant vacates the premises. Also, failure to pay rent/lease within a reasonable amount of time may turn that area back to the property owner in the form of abandonment. At that point the owner may then give consent to search and, if needed, seize evidence from the abandoned area.

The leading United States Supreme Court case dealing with abandonment of a nonmoveable structure is *Abel v. United States* (362 US 217, 4 L ed 2d 668). This case dealt with the lawful search and seizure of evidence inside a motel room that was abandoned after the occupant paid his bill and checked out. In this case, Abel was arrested in his motel room for being in this country illegally. Before leaving with the arresting officers, he was allowed to pack his personal contents, pay his bill, and check out of the hotel. After checking out, an FBI agent, with permission from the hotel management, searched the room without a warrant. During the search, evidence of espionage was discovered and seized from the wastepaper basket. The evidence lead to the arrest and conviction of Abel for the crime of espionage. The opinion of the court stated:

> No pretense is made that this search by the FBI was for any purpose other than to gather evidence of a crime, that is, evidence of petitioner's espionage. As such, however, it was entirely lawful, although undertaken without a warrant. This is so for the reason that at the time of the search petitioner had vacated the room. The hotel then had the exclusive right to its possession, and the hotel management freely gave consent that the search be made. Nor was it unlawful to seize the entire contents of the wastepaper basket, even though some of its contents had no connection with the crime. So far as the records show, petitioner had abandoned these articles. He had thrown them away. (362 US 241, 4 L ed 687)

In another court case, *Connecticut v. Zindros* (189 Conn. 228, 456 A 2d 288, abandonment was used as an exception to a warrantless search for a fire scene 11 days after the fire occurred. The premises was leased by Mr. Zindros. The Court ruled in favor of Mr. Zindros stating that abandonment was not an issue because:

1. Directly after the fire, Mr. Zindros secured the property.
2. Mr. Zindros still had personal contents within the secured premises and,
3. Mr. Zindros stated his intention was to stay at that location and reopen his business.

What is important in this case is the fact that many lease agreements state that if a fire occurs in the leased area, said area will revert back to the owner. A fire does not justify an abandonment of personal contents unless the owner of the personal contents gives up the right of ownership. Also, remember, that because the owner of personal contents has an expectation of privacy, the owner of the property cannot give consent to search, unless he or she is part owner of the personal contents in question.

As in any seizure of evidence, the governmental officials might have to justify their actions in the courtroom. The prosecutor, through the testimony of the official, will demonstrate that the lawful seizure met the requirements of abandonment. To prove abandonment, the governmental official must show:

1. Did the owner voluntarily give up all rights of ownership? Was the intent of the owner there?
2. Did the owner of the property have an expectation of privacy?
3. The location of property seized which could show a possible expectation of privacy.
4. Was the conduct of the law enforcement officer unlawful or coercive?

Chapter 5

OPEN FIELD

Another exception to the warrant involves the warrantless search and seizure, without probable cause, evidence found in an open field. The "Open Field Doctrine" is very similar to the "Plain View Doctrine" with one exception. Under "plain view," the law enforcement officer must be in a lawful position as the result of a prior, valid intrusion, before seizing any evidence. In "open field," the law enforcement officer is outside the protection of the Fourth Amendment and therefore does not have to worry whether or not the officer's position is lawful.

The United States Supreme Court first recognized the open field doctrine in the case *Hester v. United States* in 1924 (265 US 57, 68 L ed 898). The Court announced that the search and seizure upon an open field is not an invasion of privacy established by the Fourth Amendment. In the opinion of the Court, delivered by Justice Holmes, he stated:

> ...the special protection accorded by the Fourth Amendment to the people in their "persons, houses, papers, and effects" is not extended to the open field. The distinction between the latter and the house is as old as common law. (265 US 59, L ed 900)

NOTE: *Hester v. United States* is also the same case described under Abandonment.

During the year 1967, the Supreme Court again addressed the issue of the Open Field Doctrine in the case *Katz v. United States*. In the Court's discussion of "protected areas," it stated: "A private home is such an area, but an open field is not."

With this not being declared a Constitutional protected area, any evidence found either in plain view or hidden can be lawfully seized in the open field. As simple as this may sound, several court cases were brought to trial with different decisions as to what constitutes a search and seizure in an open field.

The United States Supreme Court decided to end this confusion by hearing the case *Oliver v United States* (466 US 170, 80 L ed 2d 214) and determine if a warrantless search of a marijuana field by law enforcement officers was lawful under the Open Field Doctrine.

The case involved Mr. Oliver, who was arrested by officers of the Kentucky State Police when they went to his farm on information that he was growing marijuana. To investigate the complaint, officers arrived at the farm, drove past the house and were stopped by a locked gate with a sign that stated "NO TRESPASSING." At one side of the gate was a footpath that the officers used to walk around the gate and down the road. Approximately one mile from the house, the officers found the contraband growing.

The District Court suppressed the evidence discovered by the law enforcement officers based on the fact that Mr. Oliver, by placing the "NO TRESPASSING" sign and by having a locked gate, had the expectation of privacy. Second, because the field of marijuana was growing in a secluded area, that could not be seen from any point of the public access, it was not considered a part of a search and seizure of an open field. The Sixth Court of Appeals overruled and reversed the decision stating that the marijuana field did in fact coincide with the Open Field Doctrine established in *Hester v. United States*.

The first item the United States Supreme Court looked at in *Oliver v. United States* was the decision of *Hester v. United States*. The Court concluded that the opinion announced by Justice Holmes in the Hester case was explicit enough to cover the language of the Fourth Amendment. (To review Justice Holmes' opinion see Abandonment). The Court, in affirming Hester, also added to the language that involved an individual's legitimate expectation of privacy. In the opinion of the Court, delivered by Justice Powell, he stated:

> The rule of *Hester v. United States* may be understood as providing that an individual may not legitimately demand privacy for activities conducted out of doors in fields, except in the area immediately surrounding the home. (466 US 178, 80 L ed 2d 224)

The Court further stated that the expectation of privacy expressed in *Katz v. United States* applies to an individual whose expectation must be "reasonable" by society and not an expectation demanded by an individual in the open fields.

As to the definition of open field, the Court felt that the concept can be any undeveloped, unoccupied area outside the curtilage. The field does not have to be open, nor a field by common definition. For an example, the Court used a "thickly wooded area," as coming under the open field doctrine.

The second point addressed by the Court in *Oliver* was determining the difference between an open field and the curtilage. This distinction is a fundamental element to determine if the area falls within the Fourth Amendment requirements. To identify a curtilage from an open field, the Court again reviewed Justice Holmes' opinion in *Hester* and concluded that through the common law, curtilage:

> Is the area to which extends the intimate activity associated with the sanctity of a man's home and the privacy of life...and therefore, has been considered part of the home itself, for Fourth Amendment purposes. (466 US 180, 80 L ed 2d 255)

The last subject matter addressed by the Court in *Oliver* was whether or not the law enforcement officers trespassed onto Mr. Oliver's property, thereby violating his Fourth Amendment. The Court first looked at the common law interpretation of trespass and concluded that the law was enacted to enable the owner to "exercise the right to exclude" anyone from entering the property. Trespass does not give the owner "legitimate privacy interest." Justice Powell concluded then that:

> ...in the case of open fields, the general rights of property protected by the common law of trespass have little or no relevance to the applicability of the Fourth Amendment. (466 US 182, 80 L ed 2d 227)

The final decision of the Court in Oliver stated that an open field does not have the same protection that is given to houses, papers, and effects. However, the immediate area surrounding the home or a commercial structure is called the curtilage and comes under the protection of the Fourth Amendment. Lastly, the Fourth Amendment does not apply to the common law of trespass. Based on these facts the United States affirmed the District Court's decision.

One of the disadvantages in *Oliver v. United States* and in several other lower court cases was that the search and seizure of evidence occurred in the open field and not the curtilage. Even though a separation between curtilage and the open field was made, the discussion centered around the open field.

After the Oliver case, the Supreme Court, in 1986, ruled on the Fourth Amendment issues concerning curtilage in *California v. Ciraolo* (476 US 207, 90 L ed 2d 210). Ciraolo was arrested when law enforcement officers, through a search warrant, found marijuana growing in his yard behind a six foot and a ten foot fence. The probable cause for the warrant was obtained when law enforcement officers acquired an airplane and from 1,000 feet observed, without a warrant, the marijuana plants growing in the yard. This was the only way that the plants could possibly be observed. The trial court denied motion to suppress the evidence of the search and Ciraolo pled guilty to growing marijuana. The California Court of Appeals, however, reversed on the basis that the plants were with the curtilage of the home. The warrantless observation from the airspace was then a violation of Mr. Ciraolo's Fourth Amendment rights.

The Supreme Court, in reviewing the case, reaffirmed what constitutes a curtilage from *Oliver v. United States*. Since the court established that the plants were in the curtilage, they now had to decide if the observation was unlawful. In the opinion delivered by Justice Burger, he stated:

> That the area is within the curtilage does not itself bar all police observation. The Fourth Amendment protection of the home has never been extended to require law enforcement officers to shield their eyes when passing by a home on public thoroughfares. Nor does the mere fact that an individual has taken measures to restrict some views of his activities preclude an officer's observation from a public vantage point where he has a right to be and which renders the activities clearly visible. (476 US 213, 90 L ed 2d 216)

Since the observation of the plants by the officers occurred in public airspace, an airspace that anyone could fly in, Mr. Ciraolo did not have an expectation of privacy. Therefore, the observation was outside the Fourth Amendment requirements and lawfully executed. As was stated in the Katz's decision: "What a person knowingly exposes to the public, even in his own home or office, is not a subject of the Fourth Amendment protection."

In 1987, the United States Supreme Court decided on another case in *Dunn v. United States* (480 US 294, 94 L ed 2d 326). This case dealt, again, with the Fourth Amendment and whether or not a barn was located within the curtilage of a home, and second, whether or not the expectation of privacy was violated by using the Open Field Doctrine.

Mr. Dunn was arrested by DEA (Drug Enforcement Agency) agents and convicted by a jury for manufacturing two control drugs and possession of amphetamines with the intent to distribute. The case developed when DEA agents became aware of Mr. Dunn and a codefendant buying large quantities of chemicals and manufacturing equipment for the making of drugs. Surveillance was initiated which led DEA agents to a barn located on a 198-acre ranch owned by Mr. Dunn. The complete ranch was enclosed along the perimeter by a barb wire fence and gates. Within the enclosure were other fences used to keep livestock within certain areas. The ranch house was located approximately one-half mile away from the public road. Another fence surrounded the house. Approximately 50 yards from the fence enclosing the house and not quite 60 yards from the residence were two barns. The larger of the two barns had a fence surrounding the barn with a locked gate. From the ceiling of the barn to the locked gate netting material was stretched.

Without a search warrant, a DEA agent and a local enforcement officer climbed over the perimeter fence and entered Dunn's property. Walking up to the barns, the officers smelled the odor of "what the DEA agent believe to be phenylacetic acid." Coming to the larger barn's locked gate, the officers used their flashlights to illuminate what the DEA agent believed to be a drug lab. After two more warrantless entries, the DEA agent obtained and executed a federal search warrant. Evidence was seized and Mr. Dunn arrested.

The United States District Court for the Western District of Texas denied all motions to suppress the evidence. The motions were based on the fact the search warrant was issued on information obtained during an unlawful entry. On appeal, the Court overruled the conviction stating that the barn was within the curtilage of the house. Since the house was protected by the Fourth Amendment, the Appeals Court felt that the barn did also.

In reviewing the case, the United States Supreme Court reaffirmed the Open Field Doctrines established in both *Hester v. United States* and in *Oliver v. United States*. The court confirmed that the curtilage was under the umbrella of the Fourth Amendment. The problem facing the lower courts was "defining the extent of a home's curtilage." To solve this unsettled question, the Court in *Dunn v. United States* developed a four-factor test.

1. The proximity of the area claimed to be curtilage to the home?
2. Whether the area was included within an enclosure surrounding the home?
3. The nature of the usage to which the area was put?
4. The steps taken by the resident to protect the area from observation by people passing by. (480) US 301, 94 L ed 2d 234)

According to the opinion of the court, the four-factor test did not apply to all cases involving the "extent of curtilage questions," but rather a tool to be used in the determination. Justice White stated:

> We do not suggest that combining these factors produces a finely tuned formula that, when mechanically applied, yields a "correct" answer to all extent-of-curtilage questions. Rather, these factors are useful analytical tools only to the degree that, in any given case, they bear upon the centrally relevant consideration–whether the area in question is so intimately tied to the home itself that it should be placed under the home's "umbrella" of Fourth Amendment protection. (480 US 301, 94 L ed 2d 335)

With the Court using the four-factor test in *Dunn,* it decided that the defense of Mr. Dunn did not meet the criteria involving curtilage, for the following reasons:

1. The distance of the barn to the house was 60 yards, making the barn solitary, standing by itself. Therefore, it was not "an adjunct of the house."
2. The barn, where the evidence was observed, was not within the fenced in area of the house thereby, making the barn's location outside of the home's curtilage.
3. Through information gathered by the DEA agent, the use of the barn was not part of the intimate activities in the home.
4. Mr. Dunn did little to protect the barn from being observed. By having only fences used to keep out or in, livestock, does not prevent persons from observing what lays inside the enclosed areas.

The second part of Mr. Dunn's defense, that he had an expectation of privacy was dismissed by the court. The Supreme Court felt that the position of the officers who observed contraband, did so in the open field (As determined in *Hester* and reaffirmed in *Oliver*). The court again stated that livestock fencing did not give that expectation of privacy. In Justice White's opinion he stated:

> Nothing in the record suggests that the various interior fences on respondent's property had any function other than that of the typical ranch fence; the fences were designed and constructed to corral livestock, not to prevent persons from observing what lay inside the enclosed area. (480 US 303, 94 L ed 2d 336)

The Supreme Court's decision in *Dunn* was that the law enforcement officers were standing in the open field, outside the curtilage of the ranch house making the observation of the drug lab in the barn outside of the umbrellas of the Fourth Amendment. The search and seizure, with the probable cause stemming from the open field observations, was, therefore, lawful. The Supreme Court affirmed the conviction and reversed the judgement of the Court of Appeals.

Courts have held that the area surrounding a business did not have the same intimate privacies given to the family and their home. Some areas are open to the public and, therefore, a lawful area for any law enforcement officer to enter. Those areas not open to the public enjoy certain Fourth Amendment rights with the owner having an expectation of privacy. Most commercial structures, because of their business, can expect government intrusions to conduct warrantless inspections or intrusions executed by an administrative warrant.

In the landmark case *Dow Chemical Company v. United States* (476 US 227, 90 L ed 2d 226), Dow Chemical claimed that their entire industrial complex and its buildings came under the "industrial curtilage" that covered 2,000 acres. A perimeter security was maintained to keep out the public and the view of the public from ground level. An inspector from the EPA (Environmental Protection Agency) used an aerial mapping camera without consent or a warrant to photograph two powerplants on the complex. Dow argued that as an "industrial curtilage" it should have the same protection as that given to the home's curtilage. The United States Supreme Court affirmed the Court of Appeals decision that Dow's claim of "industrial curtilage" did not meet the requirements of a curtilage found surrounding the home. In the opinion of the Court by Justice Burger, the Court agreed with the opinion of Appeals for the Sixth Circuit. Justice Burger stated:

> Viewing Dow's facility to be more like the "open field" in *Oliver v. United States* (cite omitted) than a home or an office, it held that the common law curtilage

doctrine did not apply to a large industrial complex of closed buildings connected by pipes, conduits and other exposed manufacturing equipment. (474 F
2d at 313-314)

As Justice Burger pointed out, the Court of Appeals "looked to the
peculiarly strong concepts of intimacy, personal autonomy and privacy associated with the home as the basis for curtilage protection." The
facility at Dow Chemical failed this portion of the curtilage test.

Justice Burger concurred when he delivered the Court's opinion
and stated:

> The intimate activities associated with family privacy and the home and its cur-
> tilage simply does not reach the outdoor areas or spaces between structures
> and buildings of a manufacturing plant. (476 US 236, 90 L ed 2d 236)

The one criteria the fire investigator must be aware of in the search
of the open field is to know when the open field stops and the expectation of privacy, a protected area by the Fourth Amendment, begins.

The land area between the open field and the protected structure of
the home or business is called the curtilage. This area is considered
an extension of the home and comes under the Fourth Amendment
requirements. Throughout the years, the courts have given the word
curtilage many different interpretations which depend on the facts of
each individual case, if the area to be searched is enclosed by a fence
surrounding the main structure, if the area is frequently used by the
occupants, the reasonable distance between the main structure and out
buildings and if the immediate area is off limits to the public. What
the fire investigator must keep in mind is that if the area to be searched
has a reasonable expectation of privacy, then a search warrant or a
consent must be obtained.

This knowledge of search and seizure in the open field is beneficial
to the fire investigator in determining the fire's origin and cause. Pour
patterns, trailers, containers, dropped personal property, and other
items of evidentiary value can be obtained lawfully and instanta-
neously without the scrutiny of the Fourth Amendment requirements.

Chapter 6

MIRANDA WARNINGS

There will be times when the fire investigator will respond to a possible suspicious or possible incendiary fire scene and during the course of the investigation, conduct interviews with suspects. It is just as important for the investigator to know the legality of talking to suspects as it is to be legally on the fire scene. Therefore, it is important for the fire investigator to understand Miranda Warnings, when they can be given and who can give them.

A frequent misconception made in criminal law by fire investigators from both fire and law enforcement agencies involves the use of Miranda Warnings. Do I have the authority to give Miranda Warnings? Are Miranda Warnings given before an interview is conducted with a witness to a fire? Are Miranda Warnings only given when a suspect is interviewed and/or in custody? These are some of the questions most often asked by fire investigators during the inquiry into the origin and cause of the fire. To help answer these and other questions, the development of Miranda will be discussed.

The origin of the Miranda Warnings started four hundred years ago in England when rules were formed to stop unfair methods of interrogating individuals accused of criminal activities. Admissions or confessions of the accused were obtained through torture, imprisonment, use of force, and any other form necessary to obtain the statement. Tired of the cruelty and unjust treatment, individuals began to fight, in the English Courts, to have their statements freely and voluntarily given. The lawmakers of Colonial America brought this belief from England and established it as part of the Bill of Rights in the United States Constitution. Self-incriminating evidence became a part of the Fifth Amendment, which states:

No person shall be held to answer for a capital, or otherwise infamous crime, unless on a presentment or indictment of a Grand Jury, except in cases arising in the land or naval forces, or in the militia, when in actual service in time of war or public danger; nor shall any person be subject for the same offence to be twice put in jeopardy of life or limb, nor shall be compelled in any criminal case to be a witness against himself, nor be deprived of life, liberty or property, without due process of law; nor shall private property be taken for public use without just compensation.

Self-incriminating evidence in a criminal investigation can be found in the form of confessions, admissions and exculpatory statements. A confession is a statement admitting to the necessary facts of a crime for a conviction. An admission is the acknowledgment of one or more facts that lean towards guilt, but not all of the necessary facts for a confession. An exculpatory statement is one that tends to clear a defendant from guilt. The statement can also be used to justify or excuse the defendant from guilt. Each of these three types of statements may be used interchangeably when discussing self-incriminating statements.

Confessions, admissions, and exculpatory statements may be obtained either orally, written, or both. An oral statement can be admissible in court only if the state can prove the statement occurred and was voluntarily given. A written statement is admissible if the state can show the authenticity of the document. The document does not require the defendant's signature if the state can show that the written statement was orally read to the defendant and acknowledged by the defendant to be accurate. The state must also show that a written statement was voluntarily given.

Confessions used in a conviction must be supported by other pieces of evidence. According to the book *McCormick on Evidence*, Fourth Edition, by John William Strong, a confession supported by evidence is called the "Independent Proof of the Corpus Delicti." This became necessary to stop convictions based solely on confessions that were obtained through improper police tactics and/or individuals who gave false confessions.

Normally, to show guilt of a crime the state must show three elements.
1. That the crime occurred.
2. That the event took place in a criminal manner.
3. That someone committed the crime.

For a confession to be admissible, most courts demand the evidence of the "Independent Proof of the Corpus Delicti" to be separate from the confession. This evidence not only confirms the confession but also strengthens it. In a Federal Court case, *Forte v. United States*, the court stated:

> Corroboration of an extrajudicial confession of guilt is not sufficient to support a conviction if it tends to support the confession without also embracing substantial evidence of the Corpus Delicti (68 app DC 111)

In the Federal Court case *Ercoli v. United States*, the court reaffirmed the Forte case and added that a confession:

> ....must also embrace substantial evidence touching and tending to prove each of the main elements or constituent parts of the Corpus Delicti. (76 app DC 360)

The United States Supreme Court in *Opper v. United States* (348 US 93, 99 L ed 101) verified the "Proof of the Corpus Delicti" found in Forte and Ercoli. Justice Reed, in his opinion of the Court, stated:

> The facts of the admission plus the corroborating evidence must establish all elements of the crime.

The evidence can be circumstantial or real. With a full confession, the evidence does not have to be beyond a reasonable doubt.

Up until 1964, the courts held the admissibility of an admission or confession by the "voluntariness" of the statement. If it could be shown that the interviewers or interrogators used promises, threats, or force to obtain the statement, failed to warn the arrestee of the right to remain silent, or knew the arrestee lacked educational/mental capacity, then the statement failed the test of "voluntariness." This failure made the admission or confession inadmissible in court.

The first test to confront the United States Supreme Court involving the voluntariness of a confession occurred in 1884 in the case *Hopt v. Utah* (110 US 574, 28 L ed 262). The Court's opinion, delivered by Justice Harlan, stated:

> When the confession appears to have been made either in consequence of inducements of a temporal nature held out by one in authority, touching the charged preferred, or because of a threat or promise by or in the presence of

such a person, which operating upon the fears or hopes of the accused, in reference to the charge, deprives him of that freedom of will or self control essential to make his confession voluntary within the meaning of the law.

The controlling factor as to the question of the voluntariness of a confession is found in part of the Fifth Amendment, which states: "No person shall be compelled in any criminal case to be a witness against himself."

This was reiterated in the United State Supreme Court case *Bram v. United States* (168 US 532, 42 L ed 568) which dealt with a confession that came through federal courts. Because the state courts were not binding to the federal law until 1964 (in the case *Mallory v. Hogan* [378 US 1]), the voluntariness of a confession rested on the Supreme Court case of 1936 in *Brown v. Mississippi* (297 US 278, 80 L ed 682). This case stated that a conviction within a state court based upon a confession through brutality and violence violated the defendant's due process set forth in the Fourteenth Amendment.

Again, in 1961 in the Supreme Court case *Culombe v. Connecticut* (367 US 568, 6 L ed 2d 1037), the Court addressed when a voluntary confession was admissible. Justice Frankfurter, in his opinion of the court, stated:

Is the confession the product of an essentially free and unconstrained choice by its maker? If it is, if he has willed to confess, it may be used against him. If it is not, if his will has been overborne and his capacity for self determination critically impaired, the use of his confession offends due process.

Another foundation block used in the forming of the Miranda Warnings is found in the United States Supreme Court case *Escobedo v. Illinois* (378 US 478, 12 L ed 2d 977). This case dealt with the Sixth Amendment entitled "Criminal Court Procedure." The ground work on the Court's decision was based on the section of the Sixth Amendment that states: "...the accused shall have the right to have the assistance of counsel for his defense."

This constitutional right was denied to Mr. Escobedo on several occasions, when he requested his lawyer during the four hours of police interrogations for murder. Also, his lawyer demanded numerous times to see his client, Mr. Escobedo. He, too, was denied. During the latter part of the interrogation, Escobedo made self-incriminating statements to police detectives and to an assistant Illinois state's attorney. Mr. Escobedo was then formally indicted, tried and convicted for his part in the murder. On a second hearing, the Supreme Court of Illinois held that the confession was admissible and affirmed the conviction.

In reviewing of the Escobedo case, the United States Supreme Court stated that when a criminal investigation stops being a "general inquiry" and turns toward a particular suspect, then certain steps must be taken by law enforcement personnel who detain and question the suspect. Justice Goldberg, who delivered the opinion of the Court, stated:

> We hold, therefore, that where, as here, the investigation is no longer a general inquiry into an unsolved crime but has begun to focus on a particular suspect, the suspect has been taken into police custody, the policy carry out a process of interrogations that lends itself to eliciting incriminating statements, the suspect has requested and been denied an opportunity to consult with his lawyer, and the police have not effectively warned him of his absolute constitutional right to remain silent, the accused has been denied the "assistance of counsel" in violation of the Sixth Amendment to the Constitution...and that no statement elicited by the police during the interrogation may be used against him at a criminal trial. (378 US 491, 12 L ed 2d 986)

Two years after the Escobedo Supreme Court decision, both state and federal courts were still in a legal debate over the admissibility of an individual's statement under a police custodial interrogation. To relieve the many judicial interpretations that were being decided, the United States Supreme Court heard four similar cases together. These cases, *Ernesto A. Miranda v. Arizona* (No. 759), *Michael Vignera v. New York* (No. 760), *Carl Calvin Westover v. United States* (No. 761), and *State of California v. Roy Allen Stewart* (No. 584) (now known as the Miranda Decision), were heard by the Court so that "concrete constitutional guidelines for law enforcement agencies and the courts" could be established. In each of the above cases, the defendants were detained and questioned by either a police officer, detective or a prosecuting attorney. Self-incriminating statements were obtained from the defendants while in police custody and each defendant was not given full constitutional warning of his or her rights before being questioned. As Justice Warren stated in his opinion for the Court:

> There are thus shared salient features– in–communicado interrogation of individuals in a police dominated atmosphere, resulting in self incriminating state-

ments without full warnings of constitutional rights. (384 US 445, 16 L ed 2d 707)

In the Court's long discussion in Miranda, the Justices began by upholding the Escobedo Decision as being part of the basic rights that were developed in England and made into law through the Constitution. The Court then went into length on the history of obtaining confessions in the United States from previous court cases, to the Wieckersham Commission Report on Law Observance and Enforcement, Report on Lawlessness in Law Enforcement (1931) to the Commission on Civil Rights in 1961 and finally to the contents of training manuals used by law enforcement agencies. Through these training manuals, the Court became aware that interrogation practices not only involved physical abuse but also mental or psychological abuse. Justice Warren stated: "Again we stress that the modern practice of in-custody interrogation is psychologically rather than physically oriented" (384 US 448, 16 L ed 2d 708).

In the Miranda Decision, the Court felt that the denominator for all four cases and in the Escobedo Decision was the lack of counsel as a "protective device" during the police in-custody interrogations. The Court felt that having counsel present would in effect make sure that the statement was voluntary without the use of threats, promises or use of force.

To help the suspect under custodial interrogation, the Court established safeguards to be strictly followed. They are:

1. The individual must be informed in clear and unequivocal terms, that he has the right to remain silent. (384 US 469, 16 L ed 2d 720)
2. That anything said can and will be used against the individual in court. (384 US 469, 16 L 2d 720)
3) The individual has the right to have counsel present at the interrogation. (384 US 469, 16 L ed 2d 720)
4. It is necessary to warn him not only that he has the right to consult with an attorney, but also that if he is indigent, a lawyer will be appointed to represent him. (384 US 473, 16 L ed 2d 723)

If the suspect exercises his Fifth Amendment rights after receiving the warnings, then the interrogation stops immediately. If the suspect states he or she wants an attorney present, then all questions will stop until the suspect sees the attorney and has the attorney present on any follow-up interrogations.

Based on its opinion, the Court reversed the judgement of the Miranda case (No. 759), Vignera Case (No. 760) and the Westover Case (No. 761). The Stewart Case (No. 584) was affirmed.

The main difference between Escobedo and Miranda dealt with determining when to give the suspect his or her warnings. In Escobedo, the determination centered when the suspect became the "focus of investigation," using words such as prime suspect, focal point and accusatory stage. In Miranda, the determination did not center on being a suspect, but an individual who is being questioned by law enforcement officers and is in custody or deprived in any way the freedom of movement. Words such as interrogation, custody, warning, and waivers are used.

Throughout the years after the Miranada Decision, several court cases have arisen concerning the interpretation of the words interrogation, custody, warnings, and waivers. Each of these words will be discussed, separately, through the interpretation of some of the more important court cases.

INTERROGATION

In the United States Supreme Court *Rhode Island v. Innis*, the Court came to a definition of interrogation.

> The term "interrogation" under Miranda refers not only to express questioning, but also to any words or actions on the part of the police (other than those normally attendant to arrest and custody) that the police should know are reasonably likely to elicit an incriminating response from the suspect. (446 US 291)

In 1981, in the case *United States v. Booth*, the word "interrogation" was again discussed by the courts:

> We emphasize, the test for whether questioning constitutes interrogation is whether, in light of all the circumstances, the police should have known that a question was reasonably likely to elicit an incriminating response. (669 F 2d 1231 9th cir. 1981)

In the United States Appeals Court 9th Circuit, the court determined in *United States v. Feldmen* (788 F 2d 544) that an interrogation does not exist when after an arrest the defendant's name is asked. In the 4th Circuit of the United States Court of Appeals, in the case *United States v. Taylor* (799 F 2d 126), the Court stated when a defendant invoked his right to counsel, the asking of questions by police officers concerning his identity was not an interrogation. The Court went further in the case *Pennsylvania v. Muniz* (496 US 582,110 Led 2d 528, 100 S. Ct. 2638) and stated that the gathering of biographical data on a defendant was not an interrogation and did not need Miranda Warnings to ask the questions.

Incriminating statements made to an undercover law enforcement officer does not require Miranda Warnings to be given. The United States Supreme Court, in *Hoffa v. United States* (385 US 293, 17 L ed 2d 374), stated that since the suspect did not realize the undercover officer was law enforcement, the suspect was therefore not under arrest nor under custodial care. With this belief, Miranda Warnings did not have to be given.

A suspect in custody does not have to be given his or her Miranda Warnings under exigent circumstances. If there is a threat to public safety, according to the United States Supreme Court case *New York v. Quarles* (467 US 649 81 L ed 2d 550) a police officer can, without giving Miranda Warnings, ask a suspect in custodial care questions. In the Federal Court of Appeals case *United States v. Brady* (819 F 2d 884), the court affirmed that Miranda Warnings do not have to be given when an officer questions a defendant due to the officer's concern for public safety and for the officer to obtain control of what could be a dangerous situation. A California Appeals Court, in the case *People v. Riddle* (83 Cal. App. 3d 563), set three elements to excuse the Miranda requirement in an emergency situation. They are:

1. Urgency of need in that no other course of action promises relief.
2. The possibility of saving human life by rescuing a person whose life is in danger.
3. Rescue is the primary purpose and motive of the interrogator.

Not all confessions are inadmissible without a warning. Volunteered statements given freely without compelling influences may be admissible. In the Miranda Decision, the Court expressed that "volunteered statements of any kind are not barred by the Fifth Amendment." The Miranda Decision also stated that when a suspect enters a police station and freely volunteers incriminating statements, the officer does not have to interrupt and give the required warning.

The courts have also looked at cases when officers have called suspects to voluntarily come to the police station for questioning. In *Oregon v. Mathiason* (429 US 492, 50 L ed 2d 714), the court stated that a defendant voluntarily went to the police station in response to a telephone call from an officer. The officer stated to the defendant that he was not under arrest. The defendant was taken into a room and questioned with the door closed. The court stated that the door being closed did not constitute a custodial interrogation and, therefore, Miranda Warnings were not required.

Florida's 1st District Court of Appeals, in the case *Noe v. Florida* (16 FLW D2040, 1st DCA) ruled that a woman's response to a request from a sheriff's deputy to voluntarily accompany the deputy to the sheriff's office where she was questioned without being given her Miranda Warnings, concerning the death of her three-year-old son, was admissible. During the questioning the woman blurted out that the child cried all the time and that she snapped and killed him. The interrogation stopped and the deputy then gave the woman her required warnings. The judges upheld the admission, stating that the woman was not in custody when she made the statement. The woman was not coerced and the statement was given voluntarily. The Court also stated that even though the woman was the "focus of investigation," Miranda Warnings were not required.

The main argument in United States Supreme Court, in *Beckwith v. United States* (425 US 341, 48 L ed 2d 1), was that the respondent felt that Miranda Warnings should be extended to cover an interrogation in a noncustodial environment after a police investigation has focused on the suspect. In the opinion of the Court, given by Chief Justice Burger, he stated:

> The court specifically stressed that it was the custodial nature of the interrogation which triggered the necessity for adherence to the specific requirement of its Miranda holding.

> ... Miranda specifically defined "focus" for it purposes, as questioning initiated by law enforcement officers after a person has been taken into custody or otherwise deprived of his freedom of action in any significant way.

The biggest case involving interrogation is called the Edwards Rule. This case, *Edwards v. Arizona* (451 US 477), dealt with a suspect who had been under custodial interrogation and requested an attorney. At that time, the interrogation must stop and the defendant given the opportunity to talk to an attorney before any other questions are asked. The Supreme Court stated:

> When the accused has involved his right to have counsel present during custodial interrogation...he is not subject to further interrogation by the authorities until counsel is made available to him unless the accused himself initiates further communication, exchanges or conversation with the police. (451 US 477)

This rule is important to all law enforcement officers. When in a custodial interrogation, if you are unaware that the suspect has invoked his right to counsel, any statement that is incriminating during your questioning will be inadmissible in court. Even if the defendant sees his attorney and the attorney leaves, this is not grounds to again interrogate without the presence of the attorney. The same principle applies if another officer investigating another crime interrogates the defendant after he has invoked his right to counsel. This was determined in the Supreme Court case *Arizona v. Robertson* (486 US 675, 100 L ed 2d 704). Therefore, it is important knowledge for all law enforcement officers who deal with defendants to know their history before questioning and to be aware if the defendant has evoked the right to counsel.

The defendant's initiative to speak must be in connection with the investigation. Asking for a drink of water or to use the bathroom is not a request to again start the interrogation. In the United States Supreme Court case *Oregon v. Bradshaw* (462 US 1039, 77 L ed 2d 405), the Court further expounded on the Edwards Rule. If the defendant initiates freely to again start the interrogation, the burden of proof remains for the prosecution to show that a waiver of the Fifth Amendment was established at the reinterrogation.

CUSTODY

The difference between Escobedo and Miranda is due largely to this word—custody. In Escobedo, the test was if the suspect was the "focus of investigation" before the warnings were given. However, in Miranda, the "focus of investigation" was dropped and the new test centered on custody. This was affirmed in the court case *Oregon v.*

Mathiason (429 US 495, 50 L ed 2d 719). The opinion of the court in per curiam stated:

> Any interview of one suspected of a crime by a police officer will have coercive aspects to it, simply by virtue of the fact that the police officer is part of a law enforcement system which may ultimately cause the suspect to be charged with the crime. But police officers are not required to administer Miranda warnings to everyone whom they question. Nor is the requirement of warnings imposed simply because the questioning takes place in the station house, or because the questioned person is one whom the police suspect. Miranda warnings are required only where there has been such a restriction on a person's freedom as to render him "in custody."

In another United States Supreme Court case, *Berkemer v. McCarthy* (468 US 420, 82 L ed 2d 317), the Court reaffirms belief in *Beckwith v. United States*. The opinion of the court, given by Justice Marshall, stated:

> Although the arresting officer apparently decided as soon as respondent stepped out of his car that he would be taken into custody and charged with a traffic offense, the officer never communicated his intention to respondent. A policeman's unarticulated plan has no bearing on the question whether a suspect was "in custody" at a particular time; the only relevant inquiry is how a reasonable man in the suspect's position would have understood his situation.

In the United States Supreme Court case *Stansbury v. California* (128 L ed 2d 263, 114 S Ct. 1526), the Court's opinion went into great detail concerning a suspect's Miranda Warnings when being questioned by law enforcement officers held in custody, or, not held in custody. In a unanimous decision of the court the per curiam opinion stated in part:

> ...a police officer's subjective view that the individual under questioning is a suspect, if undisclosed, does not bear upon the question whether the individual is in custody for purposes of Miranda Sec f. Inbau, J. Reid & J. Buckley, Criminal Interrogation and Confessions 232, 236, 297-298 (3d ed. 1986.)

> The same principle obtains if an officer's undisclosed assessment is that the person being questioned is not a suspect. In either instance, one cannot expect the person under interrogation to probe the officer's innermost thoughts. Save as they are communicated or otherwise manifested to the person being questioned, an officer's evolving but unarticulated suspicions do not affect the objective circumstances of an interrogation or interview, and thus cannot affect the *Miranda* custody inquiry...

An officer's knowledge or beliefs may bear upon the custody issue if they are conveyed, by word or deed, to the individual being questioned C.F. Michigan v. Chesternut, 486 US 567, 575, n. 7 (1988) (citing United States v. Mendenhall, 446 US. 544, 554 n.6 (1980) (opinion of Stewart, J.). Those beliefs are relevant only to the extent they would effect how a reasonable person in the position of the individual being questioned would gauge the breath of his or her "freedom of action." Berkemer, supra, at 440. Even a clear statement from an officer that the person under interrogation is a prime suspect is not, in itself, dispositive of the custody issue, for some suspects are free to come and go until the police decide to make an arrest. The weight and pertinence of any communications regarding the officer's degree of suspicion will depend upon the facts and circumstance of the particular case.

It is important for law enforcement officers to have explicit communication with the suspect. If the suspect believes that he or she is under custodial care, then Miranda Warnings apply. If Miranda Warnings are not given, then the suspect should be made aware that they are free to leave during any part of the questioning and that their presence with you is not an arrest. If there is any question concerning how a suspect feels about his surroundings, then give yourself this test. Ask yourself, would a reasonable man in the suspect's position understand their status involving the events that brought about the questioning?

Custody does not only apply in the police station, but also to any area where an individual is "deprived of his freedom in [a] significant way." If it is, then Miranda applies (Orozco v. Texas 394 US 324).

Once a valid Miranda Waiver is obtained from an in custody suspect, they (law enforcement officers) may continue questioning when the suspect makes an ambiguous or equivocal request for council, "maybe I should talk to a lawyer." A opinion of the United States Supreme Court in the case *Davis v United States* (114 S. Ct. 2350 (1994)) states that a clear request must be made.

Custody then applies not only to being arrested and taken to a police station for booking, but can also be at any place, at any given moment, when a law enforcement officer deprives an individual of his freedom of movement. The courts will also take into consideration if the suspect felt that his movement was restricted during the questioning. The best policy to remember is that if in doubt, read the individual his rights.

WARNINGS

The warnings are a statement that must be read to a suspect in custody before any questions are asked. These very important warnings, known as the Miranda Warnings, are important rights established in the Bill of Rights of the Constitution to protect the guilty as well as the innocent. The Miranda Warnings are given by the interrogator as follows:

I am required to warn you before you make any statement that you have the following Constitutional Rights:

- You have the right to remain silent and not answer any questions.

- Any statement you make must be freely and voluntarily given.

- You have the right to the presence and representation of a lawyer of your choice before you make any statement and during any questioning.

- If you cannot afford a lawyer, you are entitled to the presence and represetation of a court appointed lawyer before you make any statement and during any questioning.

- If at any time during the interview you do not wish to answer any questions, you are privileged to remain silent.

- I can make no threats or promises to induce you to make a statement. This must be of your own free will.

- Any statement can and will be used against you in a court of law.

WAIVER

In Miranda, a waiver is a document to present as evidence that an individual has given up his Constitutional rights. Once the warnings have been given, it is the duty of the prosecution during trial to prove the defendant understood his/her rights and voluntarily and intelligently waived those known rights. As Justice Warren stated in the Miranda Decision:

The record must show, or there must be an allegation and evidence which show, that an accused was offered counsel but intelligently and understandingly rejected the offer. Anything less is not a waiver. (384 US 475, 16 L ed 2d 724)

Just like a written consent, a written acknowledgment of the waivers, initialed by the defendant, is in most cases a show of proof the defendant was literate and knew what he was doing. Failure to sign or initial the waiver form does not make a self-incriminating statement inadmissible. According to a decision in the United States Court of Appeals Third Circuit, in *United States v. Willis*: (525 F 2d 657), the Court stated:

The fact that [accused] failed to sign the waiver of rights form is not enough, in itself, to bar admission of the confession, where other circumstances indicate that the statement was voluntarily given.

Silence, in the form of the defendant being mute as the interrogator reads each right, is not a waiver to begin the questioning. This was a concern of the Court in Miranda as stated by Justice Warren:

A valid waiver will not be presumed simply from the silence of the accused after warnings are given. (Miranda v Arizona 384 US 475, 16 L ed 2d 724)

Another concern of the Supreme Court dealt with the intelligence of the defendant. In *Brady v. United States* (397 US 742, 25 L ed 2d 747), the Court stated that waivers of constitutional rights not only must be voluntary, but must be knowing. It must be an intelligent act done with sufficient awareness of the circumstances and consequences. Mixed decisions in the lower courts have impairments such as intoxication, mental illness, retardation, or emotional stress as factors in determining if the suspect understood the warnings.

SUMMARY

Under the Fourth Amendment of the United States Constitution, individuals have the right to be secure against unreasonable searches and seizures upon themselves, their homes or their personal belongings. If a search is to be conducted by some official, it must be conducted through a search warrant or through one of the limited exceptions to the warrant. For fire investigators, the most important exceptions to the warrant are: exigent circumstances, consent to search, plain view, abandonment, and open field. As officials of governmental agencies, fire investigators must honor this Fourth Amendment right to ensure a successful criminal prosecution during a criminal investigation.

Miranda Warnings involve the Fifth and Sixth Amendments of the United States Constitution and can only be given by law enforcement officers who, one, have a suspect in some form of custody, and two, are asking said suspect questions that might illicit self-incriminating statements. Lastly, whenever in doubt, read the suspect his rights. It is better to be safe than sorry. Remember, that if your department, state, county, or municipality has rules and procedures whose scope goes beyond those found in the United States Supreme Court's decisions then abide by them.

SECTION 2

DOCUMENTING THE FIRE SCENE

WHAT IS DEMONSTRATED EVIDENCE?

SKETCHES

PHOTOGRAPHS

VIDEO

MODEL CONSTRUCTION

COMPUTER ANIMATION

INTRODUCTION

A fire investigator may be notified at any time to respond to a fire scene for the purpose of determining the fire's origin and cause. Throughout the investigation, the scene should be documented and evidence taken to confirm the determination made by the investigator. This process is necessary for two reasons. First, the information gathered will assist the investigator in writing a report, and second, the documented information and evidence taken into custody may be called upon in the future for trial in a civil or criminal court.

The best practice to develop on any fire scene is to believe that *each investigation has the potential of going to court.* Keeping this philosophy in mind, each fire scene should be properly documented so that if the fire investigator is called upon to testify, all documentation will be in its proper place. A fire investigator who uses short cuts could find his or her testimony jeopardizing the case.

Chapter 7

DEMONSTRATIVE EVIDENCE

To properly document a fire scene, three essential procedures should be performed. They are field notes, photographs, and sketching. Together, these three items form the most complete way of recording and reproducing the fire scene. Field notes, photographs, and sketching compliment one another. To omit one or more could be a disastrous and embarrassing event for the investigator testifying in court.

Field notes are only written observations made by the investigator of what he/she observes at the scene and what witnesses are saying. These notes are beneficial to the investigator as a tool to be used at a later date in writing the report and in testifying in court. When the court date arrives the investigator can only give to the trier of fact an image of the scene through narration. Without field notes, important points can be lost while testifying.

Photographs, on the other hand, do give a visual image of the scene. However, the photographs are two dimensional; they lack depth and fail the viewer in giving a true representation of distances. Shadows can blend in important features that are required for documentation. Distortion of the scene can be caused by the camera lens or the angle of the camera. Lastly, everything one sees in the viewfinder may be on the film, a conglomeration of fire debris that possibly is camouflaging evidence. A sketch includes only what the investigator believes is necessary to record. For simplicity, all minor details are left out. Photographs cannot do this. Without photographs, the fire scene cannot be reproduced in the way it was observed by the investigator for the trier of fact to view.

Sketching is another visual image that fills in the gap and supplements the field notes and photographs. Rough sketches are made from the actual presence on the scene. Finish sketches are made from the rough sketches, field notes, and photographs. Therefore, information must be accurate and identical between the notes, photographs, and sketches.

Sketches can be used by the investigator as a visual aid when talking to witnesses who can establish location of evidence, where the fire was first observed, location of witnesses in relation to the scene, and any other vital information that may be needed for the case. As a visual aid, the sketch not only helps the witnesses, but also assists the investigator in reviewing the fire scene. The sketch enables the investigator to refresh his/her memory as to the location of important items, and what the scene looked like during the investigation long after the fire scene has been destroyed.

The courtroom drama unfolds when the investigator is served a subpoena to appear at a certain date in court. The court date may be several months or even years since the investigation was completed. A good investigator feels confident going into the courtroom knowing that they have taken good field notes, photographs, and sketches.

Attorneys present their cases through witnesses and attempt to portray the character of the witness as accurate and credible. The opposing attorney's role is to disprove or discredit the opposition's witness and in so doing help defend the party they represent. During this back and forth battle, the trier of fact (in a jury trial it is the jury and in a nonjury trial, the judge) must mentally re-create what is being said and determine the weight of the testimony. Through all this verbal engagement and mental reenactment the trier of fact can become confused as to what are the actual facts in the case.

To alleviate this chaotic courtroom scene, an old concept, called Demonstrative Evidence, is finding a comeback in trial law. The major forms of this type of evidence can be found in photographs, videos, drawings, sketches, maps, models, and experiments, just to name a few. "Used as a nonverbal mode of expressing a witness' testimony," demonstrative evidence has generally been accepted to be a form of evidence not directly connected to the incident, but evidence used to illustrate or re-create a particular issue to the trier of fact. Demonstrative evidence appeals directly to the senses, especially sight. Scholars throughout the United States have written articles concerning the sensation of sight and how most individuals learn and understand better using sight more than any of the other four senses.

Because "seeing is believing" most courts want to give the trier of fact the best possible comprehension of the facts surrounding the incident. When hearing the narration of a witness these sensations are only a mental picture to the listener. Through the use of demonstrative evidence, the impression of actually being at the scene gives the trier of fact a visual contact on how the scene looked, how it may have sounded, and how it may have felt. Using demonstrative evidence in a fire investigation is especially important since the fire scene will probably, by the start of the trial, have been repaired or destroyed.

Demonstrative evidence used in the presentation of physical objects at trial have either probative or nonprobative value. Evidence that proves a particular issue as being true or false is said to be probative. The weight or importance of the evidence establishes the probative value. Nonprobative evidence does not prove or disprove an issue nor does it have probative value. Real evidence used as Demonstrative Evidence has probative value. Sketches, photographs, models, maps, computer animation, etc. used as Demonstrative Evidence have no probative value because they are used only as visual aids during the testimony of the witness.

One of the landmark cases involving demonstrative evidence occurred in 1956 in the case *W. R. Smith v. Ohio Oil Company* (10 Ill. App. 2d 67, 134 N. E. 2d 526). This case dealt in part with the use of demonstrative evidence by a medical doctor to illustrate his explanation of injuries sustained in a vehicle accident. Justice Scheineman, of the Appellate Court of Illinois, Fourth District, gave an excellent definition of the subject when he stated:

> Demonstrative evidence (a model, map, photograph, or x-ray) is distinguished from real evidence in that it has no probative value in itself, but serves merely as a visual aid to the jury in comprehending verbal testimony of a witness. Its great value lies in the human factor of understanding better what is seen that what is heard. (134 N.E. 2d 530)

In another court case, *Jackson v. State of Georgia* (255 GSA 39, 265 SE 2d 711) the court expounded on the definition of demonstrative evidence when it concluded:

Photographs, diagrams, maps, plans and similar items are generally admissible, when relevant, to describe a person, place or thing for the purpose of explaining and applying evidence and assisting the court and jury in understanding the case.

To become accepted as an exhibit in court, demonstrative evidence, such as maps, sketches, drawings, photographs, and models, must pass the admissibility test, consisting of the following:

1. It must be a correct reproduction of the scene.
2. Is nothing more than the illustrated testimony of the witness.
3. Is used to aid a witness in explaining his/her testimony.
4. It must be shown that someone knew how the evidence was reproduced and testify to the fact.
5. The evidence being presented cannot misrepresent the facts.

For the witness to testify about demonstrative evidence, the witness must show:

1. A personal knowledge of the facts.
2. That the evidence is correct or a true representation.
3. If the evidence is to be inaccurate it must be inaccurate in every detail.

To be relevant, demonstrative evidence falls under the Federal Rules of Evidence 403, that states in part:

> Although relevant, evidence may be excluded if its probative value is substantially outweighed by the danger of unfair prejudice, confusion of the issues or leading the jury, or by considerations of undue delay, waste of time, or needless presentation of cumulative evidence.

Therefore, demonstrative evidence is not admissible in all court cases. Because of its "dramatic effect," the admissibility of demonstrative evidence is determined by the discretion of the judge. In *Smith v. Ohio Oil Company*, Justice Scheineman stated:

> ...relevancy and explanatory value of demonstrative evidence is primarily within the discretion of the trial court, but, to curtail abuses, is subject to review as to the actual use made of the object. If it appears that the exhibit was used for dramatic effect, or emotional appeal, rather than factual explanation useful to the reasoning of the jury, this should be regarded as reversible error.... (134 N.E. 2d 531)

As fire investigators, the most important types of demonstrative evidence that can be utilized in presenting the case in the judicial system are maps, sketches and/or drawings, photographs, videos, and models. Each type of evidence is manufactured differently and have slightly different set of rules concerning its relevancy.

To understand the concept of maps, sketches, drawings, photographs, videos, and models as demonstrative evidence, the remainder of this section will deal with an in-depth look into each type.

Chapter 8

SKETCHES, DIAGRAMS, MAPS, ILLUSTRATIONS

This group of demonstrative evidence uses line weight as a form of pictorial reproduction that describes in detail, events, places, and the location of a person and/or object in relation to a reference point. In the early 1800s, the courts were frequently involved with confusing issues when a witness tried to verbally testify to surveys, plots and maps. The confusion diminished when the courts decided a picture of what was being verbally stated would help understand the testimony. One of these court cases occurred in the State of Louisiana in the case *Milligan's Heirs v. Hargrove* (1827, La 6 Mart NS 337), in which the court wrote:

> When witnesses give testimony how lines run, and how they intersect and interfere with each other, it is almost impossible to understand them without a diagram, which will enable those who are called on to decide, to follow with the eye the testimony that is received in relation to them.

A court in Alabama, in the case Shook v. Pate (50 Ala. 91, 1873) went further in its decision as to what constitutes a diagram:

> A diagram is simply an illustrative outline of a tract of land, or something else capable of linear projection, which is not necessarily intended to be perfectly correct or accurate...A witness may as well speak by a diagram, or linear description, when the thing may be so described, as by words. It is a common and usual method of pointing out localities and lines...

Within the drawing, a witness can use marks or shapes to represent points and objects, and lines to indicate distances, direction, and movements. The witness can develop the diagram while testifying in the courtroom or have the diagram already made prior to trial. If the witness cannot draw, the diagram can be made by someone else. However, before using the diagram as part of the witness' testimony, the witness must first testify to the correctness of the drawing and that it represents what the witness observed. The purpose of using maps, diagrams, sketches, and illustrations is to take the actual scene or event, and reduce it to a smaller version so that it may be brought into the courtroom to illustrate the testimony of the witness. Throughout early American judicial history, this was called "chalks," taken from a New England term dealing with a witness illustrating their testimony by drawing on a chalkboard.

Another reason involves the unavailability of physical evidence. Unavailable, due to the scene or objects in question being rearranged, repaired through reconstruction, lost by destroying the scene because of an existing health hazard, or the misplacement or relocation of the evidence between the event and the time of trial. Throughout the latter part of the 1800s, the courts dealt with this concept and generally agreed that:

> It is, however, a common and proper practice in courts of justice to receive models, maps, diagrams or sketches drawn on paper, or traced with chalk on a blackboard, for the purpose of giving representation of objects and places which cannot otherwise be as conveniently shown or described by the witness to the jury. (*Western Gas Construction Company v. Denver*, Ca 9th Cal, 97 F 882)

Over the years, this concept has not faltered. In 1984, in an Indiana State Supreme Court case, *Yeagley v. State* (467 NE 2d 730), the Court did not find error in the judge's decision concerning the use of a diagram that depicted the victim's house. The Court stated:

> Diagrams are admissible to represent objects and places that cannot otherwise be clearly shown or described, when offered in connection with the testimony of the witness. The diagrams that comprised the state's exhibits showed the interior layout of the victim's house. Since the house itself obviously could not be brought before the jury, those diagrams were a legitimate aid for the jury.

Maps, diagrams, sketches, and illustrations are permissible in court if the witness can testify that his or her knowledge is an accurate representation as to the physical locations, conditions, and directions observed by the witness.

One of the best statements involving this topic can be found in the court case *McCotter Transport Co. v. Hall,* (51 Del 473, 148 A 2d 110) which stated:

> It is not uncommon, and usually not improper, for counsel, in his examination of a witness, to make a sketch on a blackboard, elicit from the witness a statement that it is a fair representation of a scene or an event, and then from time to time refer to it in his examination by asking questions relative to distances or objects shown thereon, since in such cases the witness is merely using the drawing or photographs thereof as a visual or graphic means of expressing himself.

However, it must be noted that before a witness can mark or use any court exhibit, it must be properly introduced into evidence. This was addressed in the Wyoming court case, *Handford v. Cole* (402 P 2d 209) when the court stated:

> Care should be had before witnesses are permitted to characterize or illustrate their testimony by drawings or markings made in presence of jury, and markings should never be permitted until after foundational identification has been made, followed by offer and reception into evidence of exhibit.

Another important legal issue faced by the investigator is to determine whether or not the sketch or illustration is drawn to scale. If the investigator is to testify that the sketch is to scale, then everything printed in the diagram must be in direct proportion to the measurement given of the actual scene. In other words, if the scale is 1/4 inch = 1 foot, an interior wall 6 feet in length should be represented in the diagram by a line 1 1/2 inches in length. If the attorney can find fault in the investigator's scaled drawing, then the creditability of the witness can be attacked. To overcome this problem, simply inscribe on the sketch, NOT TO SCALE. If asked to testify as to what is meant by the phrase NOT TO SCALE, the reply can be something like this: "The sketch was not drawn through the use of measured proportions, but is instead a reasonably accurate and a correct detailed reproduction of the scene using approximate distances. The measurement that was taken was not precisely correct to the inch."

There have been many court cases affirming the use of nonscaled sketches. In an Oklahoma case, *Wofford v. Oklahoma* (581 P 2d 905), the court stated:

Diagram of defendant's home drawn by P.D. and verified as "fairly accurate" by the officer and defendant's wife was properly used in testimony of witnesses, although not drawn to scale.

In Montana case *Montana v. Smith*, (357 SW 2d 120) the Court stated:

A sketch of a building which the defendant had allegedly burglarized was admissible in evidence as being helpful in explaining the testimony of a witness so as to enable the jury to get a clearer understanding of the locale event though the sketch was not drawn to precise scale, since for the purpose of admission a precise scale is not important.

In 1989, a court case in Indiana, *Underwood v. State* (535 NE 2d 507), the court ruled on the concept of a diagram being not to scale. The Court stated:

Even a diagram not drawn to scale may be admitted if it is sufficiently explanatory or illustrative of relevant testimony to be of potential help to the trier of fact. The police officer testified that the diagram accurately represented the crime scene as he observed it and, unlike photographs of the scene, showed both where the body and certain property was located.

The courts have also ruled that a freehand drawing made during the testimony of a witness does not have to be to scale. In a North Carolina case, *Rogers v. State* (168 NC 112, 83 SE 161), the court stated:

In freehand drawings made by witnesses on the stand, accuracy is not possible and is not required. Consequently, the fact that the witness states that his drawing is "approximately correct" does not warrant exception to a ruling permitting its use, if it tends reasonably to illustrate his testimony in the particular for which is it used.

Simply put, maps, sketches, drawings, and illustrations are generally admissible in court if the witness is familiar with the scene depicted in the diagram, if the diagram is a fair representation of the scene and the objects within it, if the diagram is an accurate and correct representation of the scene, and the diagram is used by the witness as an aid to enable the trier of fact to understand the facts presented. The evidence must be presented in such a way as not to misrepresent the facts, inflame, or prejudice the trier of fact. This was exactly what the court stated in its decision in the case *Owensby v. State* (467 NE 2d 702, 1984, Indiana):

State's exhibit VI, which the parties have referred to as a chart, has a drawing of the room in which this incident occurred, showing generally the dimensions of the room and the placement of furniture and items within it. Both victim and a police officer testified as to the accuracy of the chart. Its purpose was to assist witnesses in articulating what they viewed so that the jury could relate the testimony to the physical surroundings and to the testimony of other witnesses. Such a chart or drawing, to be admissible, need not be a perfect representation of the subject matter. The admission of such a diagram is within the discretion of the trial court and his determination will be reversed only for an abuse of that discretion.

HOW TO DRAW A SKETCH

To formulate a sketch, the investigator does not have to be an artist to bring the fire scene and the objects together onto the diagram. By following some basic rules, nearly anyone can walk through a fire scene and prepare a sketch for future reference.

LINE WEIGHTS OF A SKETCH

One of the many problems an investigator develops when constructing a sketch is to clutter the diagram with many lines and notes of interest. What could happen at a later date, during the review of the sketch, is that the investigator will become confused to the point that it cannot be read. To avoid this confusion, different line weights should be utilized to emphasize the different areas of the sketch.

Object Lines

Is the heaviest line in the sketch that represents the edges of surfaces. In the floor plan, it is the exterior and interior walls. To distinguish the difference between the exterior wall and the interior wall, a thicker line is used for the exterior wall or a thick double parallel lines to represent the exterior walls. A softer single line can denote interior walls.

Dimension Lines

Is the next important line weight that is used to show the length of the particular surface being measured. The weight of the line is thinner and lighter than the object line. For those advanced in sketching, the numerical measurement of a room or a large area can be placed along side the corresponding surface object line. This will reduce the number of lines in the drawing.

Extension Lines

Is a fine line that relates the dimension lines to their surface. They do not touch the object lines. They start approximately 1/16th of an inch from the object line and extend approximately 1/8th of an inch beyond the dimension line.

Hidden Lines

Is used to indicate a hidden object that may be an important item to note. The line weight is medium and broken at intervals like dashes.

Break Lines

Is used to indicate that the line drawn continues, but is shown as a break. The continuation of the line is not important to the sketch.

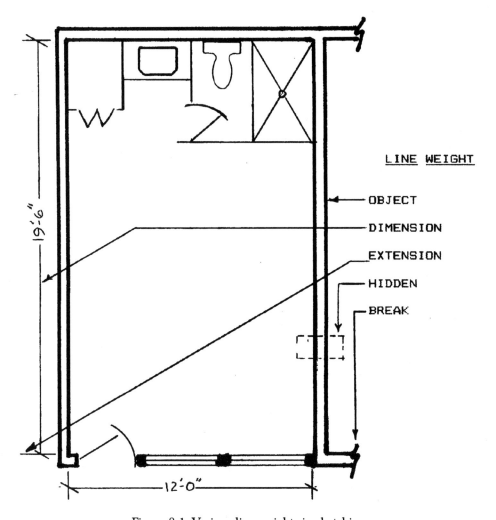

Figure 8-1. Various line weights in sketching.

If the overall sketch becomes too crowded with assorted lines and sketch notes, then it may be necessary to make several sketches of the scene in order to portray the different details. One sketch can depict the overall scene, another sketch the floor plan of the area of origin and a third sketch to show the location of evidence or any other important detail. In the case of a fatality or serious injury another sketch can reveal the location of the body.

For example, the following three sketches depict a fire scene involving a fire fatality. Due to the large area of the scene, all important doc-

umented information could not be clearly drawn in one diagram. The
scene was therefore, sketched into three different drawings.

The first drawing is a rough sketch of the complete floor plan of the
structure showing the relationship between the area of origin and the
location of the fire fatality.

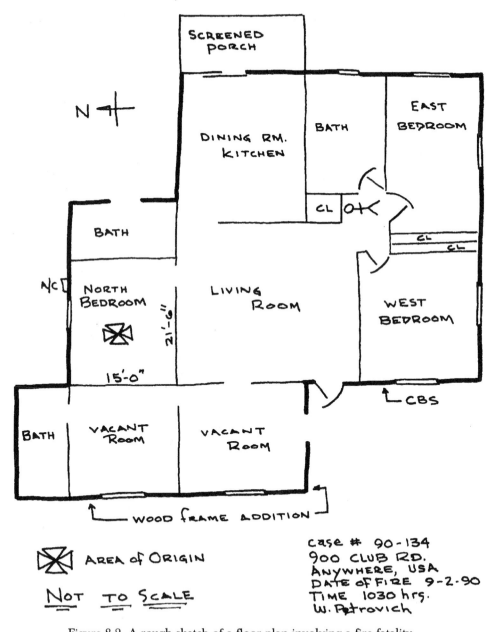

Figure 8-2. A rough sketch of a floor plan involving a fire fatality.

The second drawing is a rough, detailed sketch of the floor plan used to document numerous object within the area of origin. The fire was classified as accidental with the source of ignition being the electric hot plate being too close to combustible materials (clothing).

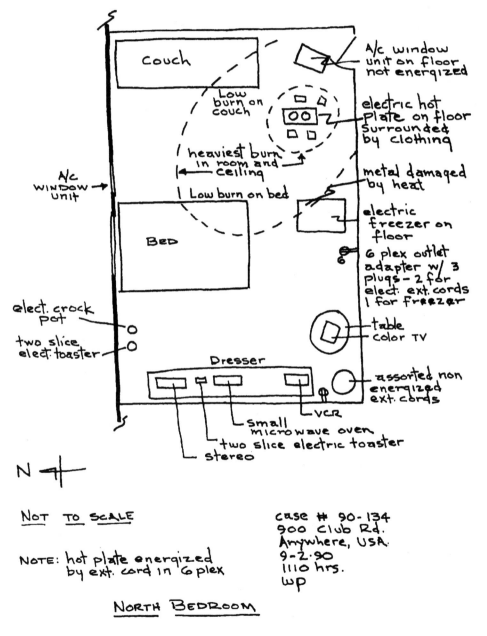

Figure 8-3. Detailed sketch of the area of origin

The third drawing details the fire fatality. This diagram also documents the measurements of the body in relation to its position in the hallway.

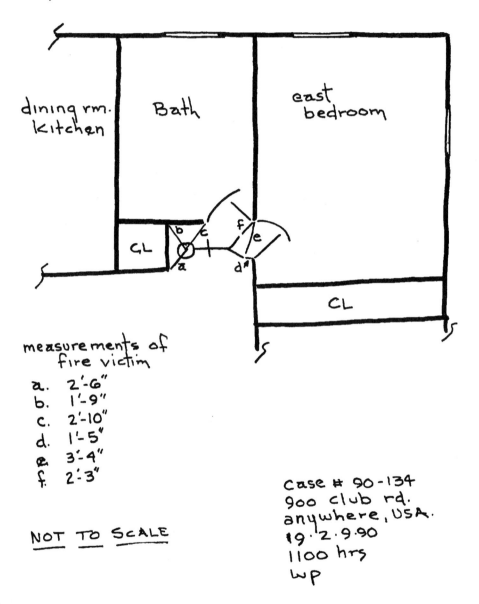

measurements of
    fire victim
a.   2'-6"
b.   1'-9"
c.   2'-10"
d.   1'-5"
e.   3'-4"
f.   2'-3"

NOT TO SCALE

Case # 90-134
900 club rd.
anywhere, USA.
19·2·9·90
1100 hrs
WP

Figure 8-4. Detailed sketch of the fire fatality.

The best test of the legibility of the sketch is to give the sketch to another individual to read and interpret correctly.

TYPES OF SKETCHES

There are five basic types of sketches that can be used at the fire scene or in the courtroom. These sketches are the neighborhood sketch, the floor plan, detailed sketch, the exploded view, and the elevation sketch.

The Neighborhood Sketch

The neighborhood sketch is used to document important items within a large scene involving a large tract of land and/or many buildings. The sketch can show the location of evidence and its relationship to the overall scene, the relationship of each building to one another, the location of major vegetation, fences, street signs, path of egress used by suspects, and the location of witnesses in relation to the scene. It is up to the discretion of the investigator to decide what should be included or excluded in the sketch.

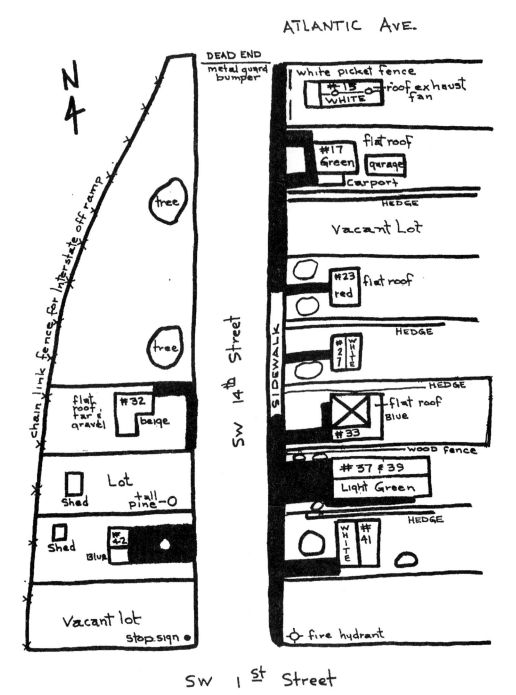

Figure 8-5. The neighborhood sketch.

The Floor Plan

The floor plan is obtained by separating the structure four feet above the floor in a horizontal sectional plane. This separation gives the viewer a bird's eye view of the location of all exterior walls as well as the location of exterior openings (windows, doors, and vents). The sketch also reveals the rooms being separated by partitions and the location of interior openings (doors, hallways, etc.)

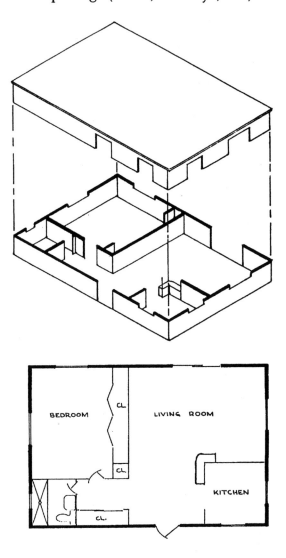

Figure 8-6. A floor plan sketch.

The Detailed Sketch

The detailed sketch shows the area of immediate concern with all the physical details that are of importance. These details can consist of pieces of physical evidence, furniture arrangements, location of doors and windows, and accurate measurements of the location of evidence. It is the decision of the investigator as to what is needed in the sketch. Figure 8-3 is an example of a detailed sketch.

The Exploded View

The exploded view is a sketch that uses the combination of the floor plan and the four elevation views (the walls) of a room. The walls and ceiling are drawn as if they are folded out with the ceiling placed on one of the walls. The purpose of this type of sketch is to show points of interest on the walls, ceiling and floor. For fire scene document-ation this is the least practical sketch to use. Trying to show heat, smoke, or fire patterns just gets too confusing to the viewer. This con-fusion is illustrated in a small bedroom fire in Figure 8-7. The area of origin is on the night stand with fire damage extending out to the bed, walls and ceiling. The exploded view is excellent to illustrate bullet holes and/or splattered blood patterns in a crime scene. This type of sketch is also known as a three-dimensional sketch or the cross-pro-jection method.

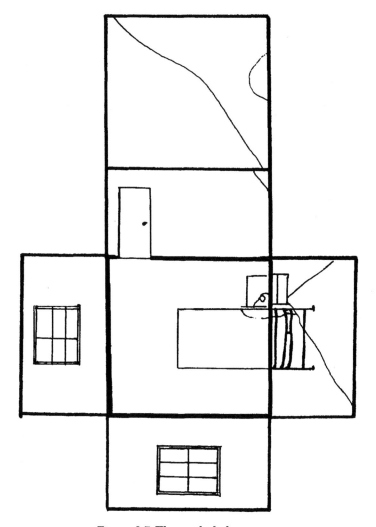

Figure 8-7. The exploded view.

The Elevation Sketch

The elevation sketch depicts either a side, front, or rear of the exterior of a structure or one of the interior walls in a room. This type of sketch is an excellent method to illustrate heat, smoke, and burn patterns as well as fire travel. Furniture and personal contents can also be documented that are found along the wall being sketched.

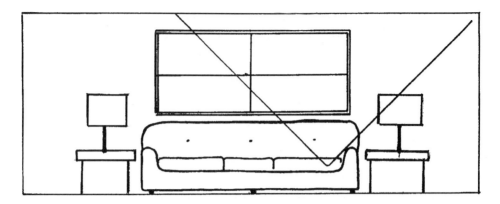

Figure 8-8. The elevation sketch.

ROUGH SKETCH

There are two kinds of sketches that a fire investigator will use in his or her investigation, the rough sketch and the finished sketch.

A rough sketch is a diagram drawn at the scene by the investigator or someone involved in the investigation. The sketch is made after a preliminary search of the scene and before any evidence is removed. The sketch should show:

- Building features (i.e., windows, doors, and how they open).
- Location of important building features (i.e., electric drop, location of propane tanks, location of gas meter, electric fuse panel boxes, etc.).
- The extent of fire, heat, and smoke patterns.
- The location of evidence (evidence that is easily destroyed has top priority in documentation and collection).
- The location of important objects, such as furniture or contents.
- Dimensions.

Dimensions of the scene are recorded first. Measurements of evidence and important objects are then recorded. Not all dimensions and measurements need to be shown. It is up to the discretion of the investigator to record only those measurements which show important distances needed to document the fire scene.

When measuring the scene, the optimum number of personnel used is three. The first position is designated as the sketcher. The second position is the tape measure reader and the third position is the holder of the tape end. Due to budget restraints in many departments, the first and third positions can be combined, enabling two people to perform the job.

Whenever possible, use a second person to verify a measurement taken. This individual can substantiate the measurement as a witness in court if needed.

TYPES OF MEASUREMENTS

When constructing a rough sketch, there are four types of measurements that can be used. They are Baseline Method, Rectangular Method, Triangulation Method, and the Grid Method.

Baseline Method

The baseline method, also known as the coordinate method, uses the principle of measuring an object between two known points. This single reference line is called the baseline. The baseline can be a wall or the known center of the room. The measurement of the evidence is taken in the following manner. A starting point is decided on one of the ends of the baseline. From that point, a measurement is made to the point that is right angle to the evidence (letter A in Figure 8-9). A second measurement is then taken from the object to the baseline (letter B in Figure 8-9). All measurements from the baseline to objects are left to right.

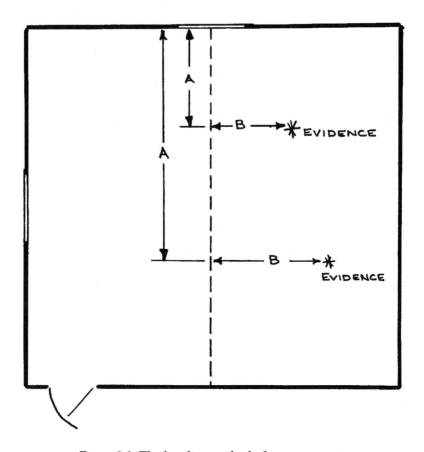

Figure 8-9. The baseline method of measurement.

Rectangular Method

In the rectangular measurement objects of evidence are located at right angle measurements from fixed baselines. In a room these fixed baselines can be the interior four walls. This type of measurement is not suited for outside measuring.

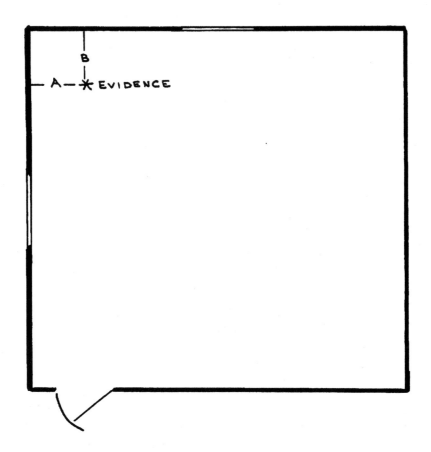

Figure 8-10. The rectangler method of measurement.

Triangulation Method

Triangulation uses three or more separate reference points that are located at the scene. From these points, a straight line is made to the evidence. A measurement is then made of the straight line. When finished, the evidence will have three measurements that will fix its location within a certain area. The reference points should be points that cannot be easily removed or destroyed. Example: telephone poles, corner of buildings or rooms, the edge of pavement.

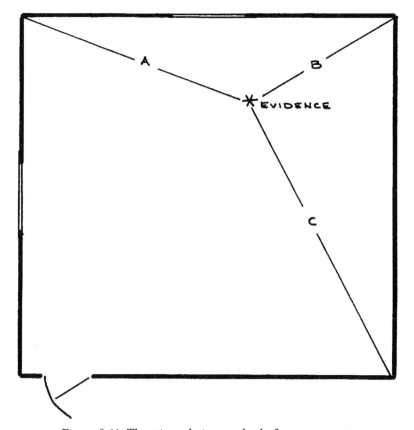

Figure 8-11. The triangulation method of measurement.

The triangulation method is especially good for outdoor scenes. Figure 8-12 shows a diagram of an outdoor fire scene involving a vehicle. For this type of measurement, the triangulation method places the vehicle and any evidence found within a prescribed area.

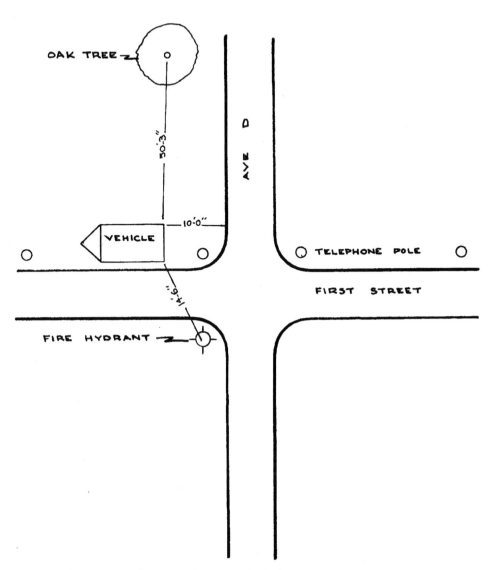

Figure 8-12. The triangulation method of measurement for outdoors.

Grid Method

This type of measurement involves the documentation of numerous pieces of evidence covering a large area. An explosion scene is a good example of when to use the grid method. Starting at a known permanent reference point at the scene, a measurement of predetermined increments (in this case every five feet) is marked off horizontally and vertically. This marking should continue at least twenty-feet beyond the furthest piece of evidence discovered. On each horizontal and vertical increment, a line or string is laid so that the scene resembles many small squares. When a piece of evidence is found, it is recorded to the nearest increment from the reference point. A measurement from the increment to the evidence is taken horizontally and vertically. For example, a piece of evidence marked number 1 on the sketch in Figure 8-13 is at 18 feet horizontal and 12 feet vertical. Evidence number two is at 28 feet horizontal and 8 feet vertical.

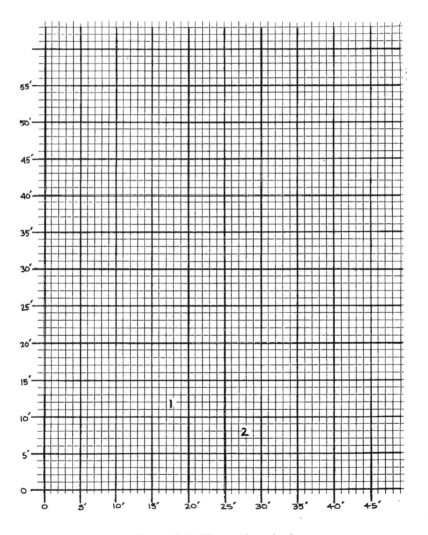

Figure 8-13. The grid method.

Once the measurements are completed, the scene and important objects are photographed, documented in field notes and, if needed, taken into evidence.

TOOLS AND MATERIAL FOR THE ROUGH SKETCH
Paper

There are many types of paper that can be used to sketch a fire scene. The type of paper to be used in drawing the sketch is up to the preference of the investigator. The inexperienced sketcher can use graph paper for sketching the scene. Even the worst sketch artist can use this type of line paper to help develop straight lines and to keep the drawing in proportion. The small blocks on the graph paper can be used to represent a unit of measure. The sketcher would determine the ratio of boxes to feet, depending on how big the sketch was to be made. For example, with a room 12 feet by 10 feet, the investigator decided to use two boxes for every foot. Therefore, the diagram would be 24 boxes horizontally and 20 boxes vertically.

Those experienced in sketching can use whatever paper they are comfortable using. Plain white 8 1/2" x 11" paper or notebook paper are preferred, both for ease of use on the scene and the way they fit in the file folder. Sketches should never be drawn on scrap pieces of paper. This type of paper can be easily lost and if it becomes an exhibit of the court, will show a lack of professionalism.

Pencils

Pencils are best used so that mistakes can be erased easily. The ink of pens leaves messy blackened-out lines in a mistake. At least two different types of line weights should be used for clarity.

Symbols

A uniform set of symbols should be developed before the first sketch is drawn. Architects throughout the United States have a uniform set of symbols that represent different construction segments of a structure. These same symbols can be incorporated in the fire scene sketch.

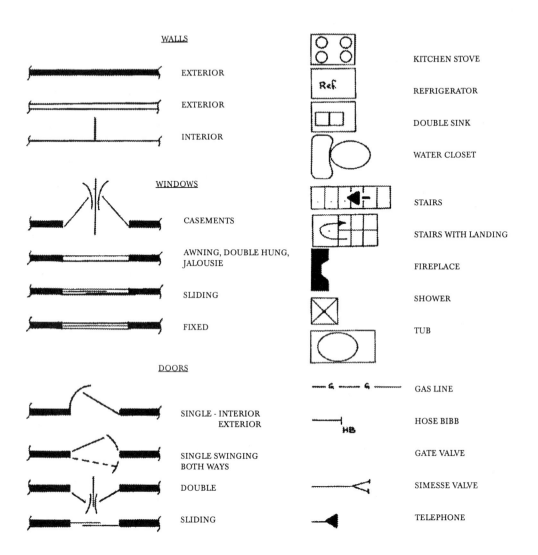

Figure 8-14. Basic symbols to identify important features in a structure.

ELECTRICAL SYMBOLS

	DUPLEX OUTLET
	DUPLEX OUTLET FOR GROUND
	WATERPROOF OUTLET
	FOURPLEX OUTLET
	COMBINATION SWITCH AND OUTLET
	LAMP AND PULL SWITCH
	FLOOR OUTLET
	JUNCTION BOX
	SPECIAL PURPOSE OUTLET
	FAN OUTLET
	CEILING OUTLET FOR RECESSED FIXTURE
	FLUORESCENT LIGHTING
	WATER HEATER
	WATER PUMP
	SERVICE PANEL
	ELECTRIC METER
	ELECTRIC METER
	AIR CONDITIONER
	A/C RETURN AIR DUCT
	SMOKE DETECTOR

Figure 8-14. Basic symbols to identify important features in a structure.

Fire scenes not only include structures, but also vehicles (automobiles, trucks, and heavy equipment), water vessels of all sizes and wild lands. For the automobile, a predrawing can be constructed depicting the different exposures. At the fire scene the investigator can use the predrawing to indicate observed fire and heat damage, and the location of possible evidence. If a rough sketch is made without the predrawing, a rectangle can be drawn to document a vehicle fire. The rectangle represents the outline of the vehicle. A triangle is placed on either end of the rectangle and is used to indicate the front of the vehicle. The rectangle is divided into three sections to indicate the engine, passenger and trunk compartments. This type of sketch can also be used for trucks and boats of all sizes.

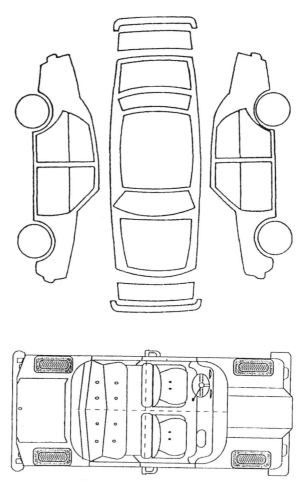

Figure 8-15. Predrawings to document vehicle fire damage and location of evidence.

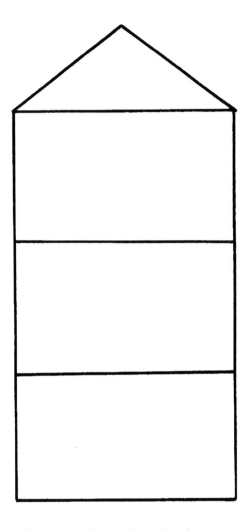

Figure 8-16. A rough sketch to indicate fire damage in a vehicle.

Once the investigator has completed the rough sketch and has left the scene, the sketch should not be altered in any way. Keep the rough sketch in the work file. It may be required as an exhibit of the court if the finished sketch is found to contain inconsistencies.

RULES FOR DRAWING THE ROUGH SKETCH

1. Decide what is to be sketched.
2. Determine the compass directions and indicate north on the sketch.
3. Control all measurements.
4. Have another officer verify all measurements.
5. Don't estimate the distance by pacing or by foot steps.
6. Use a measuring tape or ruler—be accurate.
7. Locate all objects accurately.
8. Include all essential items in the drawing, exclude the irrelevant.
9. All sketch corrections must be made at the crime scene.
10. When locating the position of a body, two separate sets of measurements are taken: One set of measurements is made from the head, the other from the feet. It is also helpful to document the location and position of the extremities.
11. All bodies drawn on a rough sketch are shown as stick figures.
12. All objects in a rough sketch are identified either with numerals or letters.
13. All objects are described in detail (size and position) in the notes—don't clutter the sketch.
14. Include the following information in the notes: date, time, case number, location of incident, weather conditions, light, names of the victim and of the officers present, etc.

FINISHED SKETCH

The finished sketch (see Figures 8-19 & 20) is a well-prepared diagram using instruments for the construction of the lines. The purpose of the finished sketch is to bring the fire scene into the courtroom for presentation. It can strengthen the testimony of the witness by showing a neat and accurate drawing. This can create a favorable impression with the trier of fact. It also gives the witness a respect for their efforts and the ability to present the facts in a clear and concise manner. A finished sketch does not have to be drawn by the investigator involved in the making of the rough sketch. However, the investigator who made the rough sketch must verify in court to the accuracy of the finished sketch.

To help prepare the finished sketch, architect drawings from builders, architects, building departments, city engineering departments, and the inspection reports from the bureau of fire prevention, to name a few, can add information about the fire scene. Also, look for remodeling plans that may show recent work. Combining these drawings with the rough sketch, photographs, and field notes the investigator should have a complete and accurate finished sketch.

TOOLS AND MATERIALS OF THE FINISHED SKETCH

Before beginning the finished sketch, the investigator should decide on the purpose of the drawing. Is the sketch going to be a part of the official report, or is the sketch going to be a part of the courtroom testimony? If used in the courtroom, then a decision must be made as to how large the sketch is going to be.

Paper

If the finished sketch is to be a part of a report, a clean 8"x 10" or a legal size white piece of paper can be used. For courtroom presentation, the size will be up to the discretion of the investigator. Remember that the sketch will be reviewed by the trier of fact. With this in mind, the larger the sketch the better the clarity will be. There are assorted sizes of white poster board to choose from. The disadvantage of using poster board is that it is a thin piece of cardboard and not very rigid. To display the poster board sketch, fasten the board onto an easel. Without the fastener, it may become embarrassing when the sketch falls from view during the testimony.

Another type of poster board, called foam board, is better to use than regular poster board. Foam board is very rigid, constructed of an 1/8 inch foam sandwiched between two pieces of poster board. This board also comes in various sizes.

Pens-Pencils

Initial finished sketches should be drawn and completed in pencil. This procedure assures the drawer that all structural lines are correct and properly related to the fire scene, rough sketch, photographs and other forms of information that is being used. Once completed in pencil, the different lines weights in ink will give the sketch a professional look.

Symbols

The same architectural symbols used in the rough sketch may be incorporated in the finish sketch.

Tools For Making Lines

In order to construct the uniform, straight lines of the finished sketch, several different instruments are required.

Rulers

There are different types of rulers that can be used—the architect ruler, the engineer ruler, and the standard 12-inch ruler. Not only does the ruler measure, but can also be used for straight line construction.

T-Square

Used to draw horizontal lines and is the base for the 90 and 45 degree triangles.

Triangles

Used to make vertical lines and degree angle drawings.

Templates

Are assorted plastic sheets constructed with punched out forms that are used to make various sizes of circles, squares, curves, room furniture, crime scenes, just to name a few. These are not required, but can be very helpful to the sketcher.

Drawing Board

A large board of various sizes used to tape the paper onto the surface. The T square lines up on the side of the drawing board and is used to make horizontal straight lines. The T square is also used in conjunction with the 45 and 90 degree angles to make vertical straight lines. The drawing board is also a hard foundation to construct the lines of the sketch.

THE COMPUTER SKETCH

Before the age of computers a finished sketch was drawn by hand. Today, computer sketches can be readily constructed using a variety of software programs. New advancement in computer technology has made the images of the computer sketch sharper. The lines are clearer and the letters uniformed. Technology has also made the task of the computer sketcher easier through the use of a CAD system. CAD enables the sketcher to determine the scale, pick out premade construction features, furniture, and household items and place them in the drawing. (For further information on computers refer to Chapter 12 "Computer Animation.")

To show the differences between rough and finished sketches, an actual fire scene sketch will be used. The first drawing is the rough sketch that was drawn at the fire scene during the fire investigation. The second drawing is a finished sketch that was hand drawn using the rough sketch, photographs and field notes to complete. The third drawing is another type of finished sketch constructed through the use of the computer.

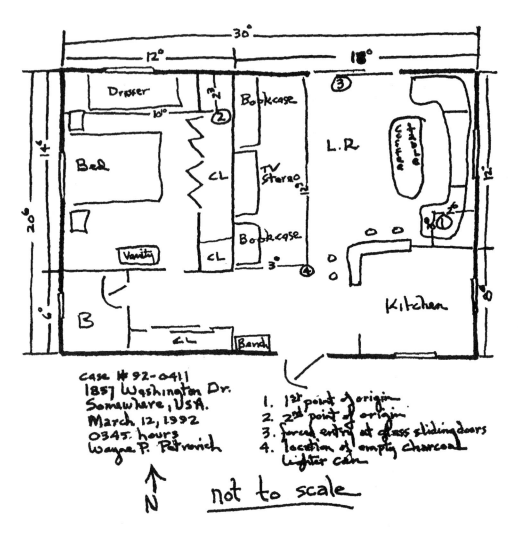

Figure 8-17. A rough sketch of an actual fire scene.

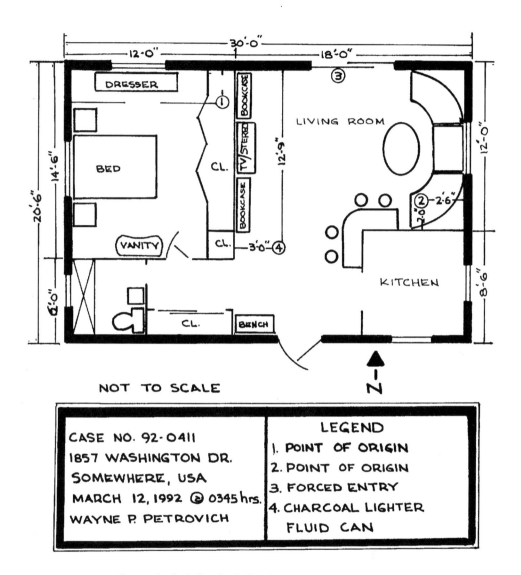

Figure 8-18. A finished sketch of an actual fire scene.

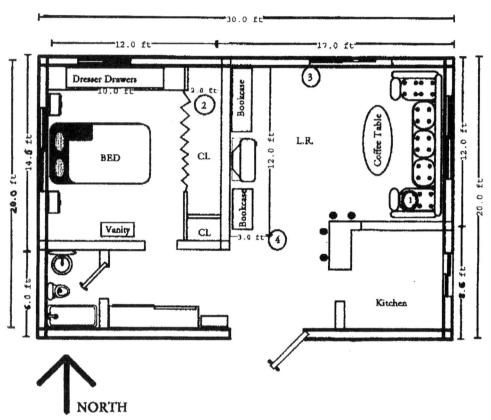

NORTH

Case number VF 92-0411
1857 Washington Dr.
Somewhere, USA
March 12, 1992
0345 hours
Wayne P. Petrovich

1. 1st point of origin
2. 2nd point of origin
3. Forced entry at glass sliding door
4. Location of empty charcoal lighter can

NOT TO SCALE

Figure 8-19. A finished, computer sketch of an actual fire scene.

When recording either the rough or finished sketch, several important features must be included:

1. The investigator's name.
2. Date and time of the fire.
3. Address of the fire scene
4. Date the sketch was made.
5. Name of the person who made the sketch.
6. Who made the measurements and who assisted.
7. For orientation purposes, the direction of NORTH should be denoted by the word NORTH or the letter N with an arrow indicating the direction.
8. Type of scale or if none indicate it with the phrase NOT TO SCALE.
9. A legend, which is an explanatory list of items in the sketch with letters of the alphabet to denote furniture and fixed items and numbers to indicate pieces of evidence.

Chapter 9

PHOTOGRAPHS

Photographs are one of the most utilized forms of Demonstrative Evidence used in the courtroom today. As a visual reproduction, photographs, through pictures of objects, scenes, or people, tell a story of an action or event that is preserved within a moment of time. As fire investigators, it is just as important to document an accidental fire as it is to prove that the crime of arson has been committed. Thus, photographs of the fire scene help establish the points of proof of the fire's origin and cause.

Photographs will record everything within the view of the lens. The pictures will be unbiased with a high degree of reliability. In the courtroom, photographs can incite passion and prejudices that the testimony of a witness cannot. The witness can forget or distort what was observed at the scene. Photographs thus become an illustration of the testimony of the witness.

The word "photograph" comes from two Greek words: "Phos," meaning light and "graphos," meaning write. Best translated, the word photograph means to "write with light."

We tend to think of photography as a technical invention created in the twentieth century. However, the first camera was developed some time in the seventeenth century and was called a "camera obscura." This type of camera consisted of a light tight box with a mirror, some type of screen, and a pinhole or a crude lens. The "camera obscura" was used by the painters to bring in a large three-dimensional scene into a small two-dimensional scene implanted on the screen. The picture would remain as long as the "camera obscura" remained in place.

The start of modern photography began in 1826, when a French chemist, Joseph Nicephore Niepce, created the first authentic photograph. The picture was an eight-hour exposure of Niepce's courtyard. Niepce then became partners with a French painter by the name of Louis Jacques Mande' Daguerre. Together they tried assorted combinations of metal solutions and surfaces where a picture would remain permanently. After a few years the partnership dissolved. Using a camera with lenses made by Daguerre and the different processes learned with Niepce, Daguerre accidentally discovered the fumes of mercury impinging on an iodized silver plate that was exposed to a scene. The image permanently remained on the plate. The pictures became known as Daguerreotypes.

The first court case to use a photograph as evidence occurred in November of 1839 during a divorce hearing when the husband photographed his wife having a meeting with her lover. Also in France, the first "mug shots" were taken in 1841. In the *Philadelphia Public Ledger* of November 30, 1841, an article was written that stated the police in France were using daguerreotypes to identify criminals. The article read in part:

> When a discovery has been made in science there is no telling at the time to what useful purpose it may afterwards be applied. The beautiful process invented by Daguerre, of painting with sunbeams, has been recently applied to aid the police in suppressing crime. When any suspicious person or criminal is arrested in France, the officers have him immediately daguerreotyped and he is likewise placed in the criminal cabinet for future reference....

In 1859, the United States Supreme Court heard the first case *Luco v. United States* (64 US 515, 16 L ed 545) involving photographs that compared the authenticity of signatures in a civil case. The first criminal case to use photographs, *Udderzook v. Commonwealth* (76 Pa St. 340), occurred in the State of Pennsylvania in 1874. Photographs taken while the victim was alive were used to identify the deceased victim in a murder case. In part, the court's opinion stated:

> That a portrait or a miniature, painted from life and proved to resemble the person, may be used to identify him, cannot be doubted, though, like all other evidences of identity, it is open to disproof or doubt and must be determined by the jury. There seems to be no reason why a photograph, proved to be taken from life and to resemble the person photographed, should not fill the same measure of evidence.
>
> The daguerrean process was first given to the world in 1839. It was soon followed by photography, of which we have had nearly a generation's experi-

ence. It has become a customary and a common mode of taking and preserving views as well as the likenesses of persons, and has obtained universal assent to the correctness of its delineation.

Because of this view concerning daguerreotypes (photographs), the opinion of the court in Udderzook has been the foundation of admitting photographs into court throughout the years. In Udderzook, the opinion of the court stated in part:

> We know that its principles are derived from science; that the images on the plate, made by the rays of light through the camera, are dependent on the same laws which produce the images of outward forms upon the retina through the lenses of the eye.
> The process has become one in general use, so common that we cannot refuse to take judicial cognizance of it as a proper means of producing correct likenesses.

Towards the end of the 1800s, photographs came under the same rules of evidence as maps, sketches, and diagrams. If the photograph could be shown to be a fair representation of the subject in question, then the photograph was admitted into evidence as a means to help the witness explain his or her testimony. When copyright laws started dealing with the concept of photographs and motion pictures in the early 1900s, the federal courts combined these new terms into the definition of "writings." As the Uniform Rules of Evidence, Rule #1, Subsection 13 states:

> "Writings" as handwriting, typewriting, printing, photostating, photographing and every other means of recordings upon any tangible thing, any form of communication or representation, including letters, words, pictures, sounds of symbols or combinations thereof.

The two important photographic inventions introduced into the United States that helped in fire investigation were the photo flashbulb and color film. In 1930, photo flashbulbs made possible the picture taking of evidence in dark or semidark areas. Prior to this, flash pictures were obtained by using flash powders that were explosives, leaving a large volume of smoke.

In 1943, the first appellate court case, *Green v. City and County of Denver* (142 P 2d 277, 111 Colo. 390), became the first case to use color film as photographs evidence. The photographs taken were of spoiled meat being sold to the public in violation of health codes.

The use of black and white film will show high and low colors of grey, black and white, thus distorting some photographs. Especially in the grey and black environment of the fire scene, evidence can be camouflaged on the black and white film. With the introduction of color film, the detail enhances the quality and produces more realistic view, as seen at the time of the investigation.

THE MODERN CAMERA

The basic principles on how a camera operates is the same principle as the human eye. Both use light that is reflected from a subject and is projected onto a light-sensitive surface.

Light from the object is reflected to the eye. The amount of light brightness entering the pupil is controlled by the iris. From the iris the light passes through the lens and is focused onto the retina upside-down. The retina converts the light to nerve impulses, corrects the image and sends to the brain the sensation of sight.

Light is reflected from the object and enters the camera through the lens and is focused. From the lens, the light passes through and is controlled by the aperture ring (f-stop) and the shutter speed (time interval the light is allowed to hit the sensitive surface).

The aperture ring and the shutter acts like the pupil and iris of the eye. Instead of the focused image impinging on the retina of the eye, the light is hitting the film in the camera.

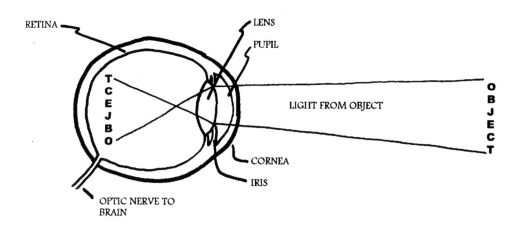

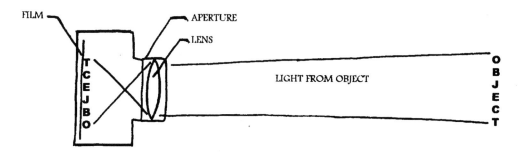

Figure 9-1. Comparison between the camera and the human eye.

To fully understand the complete operation of a particular camera, consult the instructions found in the operating manual.

The needs of the fire investigator will determine the camera that will be used. With today's sophisticated and complex electronic printed circuits, automatic cameras can be damaged by physical shock through rough handling, dropping onto hard surfaces, and from the harsh environment of the fire scene which can consist of water, humidity, heat, cold, fire debris, smoke, and acid gases, just to name a few.

The first thing then is to determine if the camera chosen can stand up to the punishment of the fire scene.

Second, before investing money on a camera and its support equipment, find a similar model that is interesting to you and try it out. Use the camera in different positions, in different types of lighting and become familiar with the different control mechanisms. Remember, the camera should be compact, light in weight, easy to use, dependable and durable. Only by actually using the camera will the investigator develop its potential and determine what camera is best.

WHICH CAMERA IS FOR YOU?

On the market today, assorted cameras in various sizes and functions may be purchased. The choice made by the investigator or the department will depend on the department's budget and how involved the investigator wants to become in taking the photographs. On one end, the camera is inexpensive and all the investigator has to do is point and shoot. At the other end, the electronic circuit board camera is expensive, with options of elaborate accessories and a basic knowledge of photography needed to operate. In most instances, the lens of the expensive cameras are of a better quality and the bigger film format produces a sharper, detailed, and color saturated picture. With so many cameras to chose from, a happy medium can be reached somewhere between the two extremes.

To understand the different cameras available, terminology such as format, viewfinder, range finder, single lens reflex (SLR), fixed focus, variable focus, auto focus and interchangeable lens will be examined.

The format of the film determines the camera's size; more precisely, it is the size of the negative. There are four popular sizes now being used. They are the 110 for pocket cameras, the 126 for larger cameras, the 35mm for the viewfinder and SLR cameras and last, the 120-620, better known as the 2 1/4 inch square format, used in SLR and twin lens reflex.

The viewing system of the camera is one of two designs. The first design is called a direct viewfinder, where a separate view of the subject is observed through a window in the camera's body instead of the lens. The photographer should be aware that the view finder is an approximation of what the camera lens sees and a framing error may occur. Found in better quality viewfinders, the range finder is used to help focus the lens. Through another window, separate from the viewfinder and the lens, the range finder contains mirrors that give a second image of the subject in the viewfinder. To become focused, the double image will come together as a single image when the correct focus is obtained.

The second design is called reflex or SLR (single lens reflex). This design of mirrors lets the photographer see the subject through the same lens that the camera uses to project the subject onto the film.

A camera will have one of four basic types of focusing capabilities. In the first one, called a fixed focus, the camera's lens is immovable and the aperture is very limited or fixed. This type of focusing requires little training. The lens keeps a fixed focus on the subject from approximately six feet to infinity. Basically, all the photographer has to do is point the camera and shoot. This camera is usually the least expensive to buy. The second type is called a fixed lens with variable focus. This type of camera has a fixed normal lens with a focus ring. Other cameras may have both the focus ring and the aperture ring. The focus ring may be done manually thorough a distance scale or through symbols. Some knowledge is needed in the use of the aperture, shutter and depth of field. The third type of focusing is called autofocus. Through the electronic circuits the camera determines the distance between the camera and subject and then through a tiny electric motor moves the lens for focusing, automatically. With this feature the photographer needs little knowledge of photography. The autofocus is a better quality point and shoot camera. The last type of focusing camera has a special feature where the lens can be disconnected from the camera's body and replaced by another lens. This is called interchangeable lens. The most used interchangeable lenses are:

1. *Normal lens* –should be the standard used in photographing evidence. The angle of view is around 50 degrees, which is approximately the zone of sharp focus of the human eye with out moving the head.

   When a photograph, taken with this lens, is made into an 8 inch x 10 inch enlargement and held out at approximately 15 inches from the eyes, it will produce a realistic prospective of what the normal eyes observed of the scene.

2. *Macro lens* –used for close up work to bring out small details.
3. *Telephoto*–used to bring the scene closer to the photographer without moving in. Can be used for surveillance.
4. *Wide angle* –used in close quarters when an overview is needed. When using a wide angle lens, be careful of the distortions that will be produced.

TYPES OF CAMERA

There are six (6) different types of cameras according to the viewing system and format. Keep in mind that in this day and age of modern technology, new concepts and designs are constantly being made that help upgrade existing systems.

The 110 mm Pocket Camera

The 110 mm pocket camera (also known as the 110 snapshot). This camera is the least expensive to buy. Most cameras have a viewfinder, fixed focus and f-stop. Some of the newer models have limited automatic exposure control. Most have a built-in flash with limited distances. They are lightweight, slim, and easy to use–a true point and shoot camera.

The 126 Camera

The 126 camera is a lightweight viewfinder camera with a bigger format. The majority of the cameras have a fixed focus and f-stop. More expensive models may have range finders, autofocus, and a built-in flash. This is another lightweight and easy–to–use point and shoot camera.

The 35 mm Viewfinder Camera

The 35 mm viewfinder camera is a fixed lens camera with the more expensive cameras having a range finder, a dual system of lenses providing wide angle and telephoto capabilities. Most of the 35 mm viewfinder cameras will have some or all of the following features:

Built-in flash, automatic exposure, automatic focusing and built-in light meters. Because this type of camera is a 35 mm, the many different types of films available for the SLR can also be used. The snapshot 35 mm is a lighter, cheaper, and simpler to use than the SLR, making it an excellent camera to carry as a backup.

The 35 mm SLR Camera

The 35 mm SLR (single lens reflex) camera has interchangeable lenses with a viewing system that lets you see the image exactly what is being projected on the film. The 35 mm SLR equipment accessories vary according to the needs of the photographer. Besides the lenses, the camera may use different filters, flashes, film speed, and techniques to produce a photograph. Another advantage of the 35 mm is that it can use film with more exposures than other types of cameras. This camera is lightweight, easy to handle and put away for future use. The camera can withstand a medium amount of abuse. This versatile camera can meet the requirements and demands of the fire investigator.

Twin Lens Reflex

The Twin lens reflex has a large negative that is a 2 1/4 inch square. The large negative gives a sharper picture. This type of camera is good for landscape and portrait work. The viewing system is located at the top of the camera and should be held at waist level for the best results. The number of exposures on a roll of film will not exceed 12. The twin lens reflex is not a suitable camera to use in fire scene investigation work.

The Instant Camera

The Instant camera, better known as the Polaroid®, makes available a finished print of an object within seconds of taking the photograph. Polaroids should be used as a backup to insure a photograph is taken of important items or in a fatality. After taking normal photographs of the fire scene, follow up with a few shots from a Polaroid. This can keep the integrity of the fire scene secured if:

a. There are problems in film development.
b. There is equipment failure.
c. There is operation malfunction.
d. The roll of film is lost.
e. There is not enough time to set up with first line camera equipment before evidence is destroyed.

FILM

Once the camera brings the object into view, it must place the image onto a permanent surface called film. In simple terms, film is a transparent plastic base that is coated with a light sensitive emulsion. For black and white film the single layer of emulsion contains silver salts called halides suspended in a gelation. In color film there are three different layers of emulsions and dyes, each sensitive to a different color. Once the film has been exposed to the image it is placed in a chemical solution. A chemical reaction then takes place causing the exposed silver salts to become grains of black metallic silver.

The size and amount of grain determines the film speed. A slow speed has a grain texture that is even and small. A faster film speed has a bigger and more pronounced grain. A numerical value is given to the film indicating its sensitivity to light. In the United States this number is called the ASA (American Standards Association). The higher the number, the faster the film speed. A slow speed is a rating of ASA 25 to 64. This film can be used in bright or average light. The photograph of normal size will not show the grain. Even most enlargements will be free of the grain. A medium-rated speed, ASA 64 to 125, can be used in average light conditions with minimum grain in the photograph. A fast-rated film speed, ASA 125 to 400, is used for below average lighting. At this speed, grain becomes noticeable, especially in the 8 inch by 10 inch enlargement and up. With the help of modern technology, a new ultrafast film speed has been developed. This speed, ASA 400 to 1600, can be used in very low light. Expect to see grain in the photograph.

Another standard of film speed often seen on film and cameras is the DIN (Deutsche Industrie Norm). Developed by the Germans, the concept is the same as the ASA except for the numbering system. The numbers will run from 13 to 33. The higher the number, the faster the speed. A recent development was made to standardize film speed worldwide through the ISO (International Standards Organization). The numerical values of this system correspond to the scale values of the ASA.

The fire investigator must decide, before photographing the fire scene, on what format of film to use: prints, slides, or instant. Each format has a specialty with advantages and disadvantages to their use.

Print film produces a negative film that can reproduce the photograph onto unlimited amounts of prints of various sizes. The negative film projects its image onto color printing paper by projecting a light through the negative. The positive print is then produced by placing the exposed paper into a chemical bath. If the exposure of the negative is slightly over or underexposed, then the correction can be made during the printing. The color quality is not as good as slides due to the print paper absorbing part of the light. Another reason in the reduction of color quality is found in the faster speed film's larger grain. Another advantage of print film is the use of contact sheets. By placing the exposed negatives on printed paper, the result is the same size positive prints. The contact sheet can be used to determine what frame to print or enlarge. Today, with limited budgets in government spending, police and fire departments may find contact sheets an economical way of storing and preserving printed photographs.

Slides, also known as transparencies, produce a positive color image on the film base. Each frame of the film is individually mounted in a cardboard or plastic border. The print is small and hard to observe with the naked eye. A hand viewer or a slide projector magnifies the image into a brilliant and sharp color photograph.

When previewing the slide, be sure the photograph is in the projector correctly. It can be embarrassing and possibly inadmissible if shown incorrectly during your courtroom testimony. Check for the proper sequence; if the frame is in backwards or upsidedown, correct it during the preview. Slides are economical and easy to store. The main disadvantage of slides centers around the use of the projector in the courtroom. Mechanical breakdown or a light bulb malfunction may occur. In a projector, slides can only be shown one at a time. When two slides are to be compared, then a second projector is needed. If a large amount of slides is presented, one at a time, the time interval will be long and the narration from the first photographs forgotten. When the trier of fact wants to review particular photograph slides, a long procedure would take place to find the appropriate slides. If at all possible when using slides in trial, have a set of prints made from the slides. Position the prints in the same sequence as the slides. This procedure may come in handy during the deliberation of the jury if they request to see the photographs again.

Instant film, used in the Polaroid cameras, consists of what is called "peel apart" film. When the shot is taken the exposure will come out of the camera. After 60 seconds, the negative is peeled apart from the positive print. The negative is then thrown away. There is no choice as to the film speed. Instant film may be reprinted and/or enlarged, but it is an expense not suited for fire or law enforcement budgets. The print is an odd format, usually 4" x 4," with the enlargement either 6" x 6" or 8" x 8."

Regardless of what type of film is used, the negative, slide or instant, it must be handled carefully. If they are lost or damaged, so also may be the case.

Once the brand of film has been decided, stay with that one particular brand. Brands can change in color variations and differ in color performance under different lighting. Most film brands have various film speeds and each will differ in grain and color contrast.

To get the best out of any type of film, use it within the date stamped on the package. Store the film in a cool dry environment. For long-term storage, it is recommended to store the film in a sealed container placed in the refrigerator.

DOCUMENTATION OF THE FIRE SCENE WITH PHOTOGRAPHS

To document a fire scene with photography, both the exterior and interior should be photographed as soon as possible to show the nature and extent of damages. The fire scene, as with any other crime scene, should not be disturbed after discovery until photographs and field notes are taken.

Photo Log

A photo log should be used to help keep an accurate record of the photographs. Without this log, documentation of the photographs would be through memory. The photographer must be able to explain in court what each photo depicts, its location to the overall scene, and why the photo was taken.

The purpose of the photo log is to record and describe what is in each exposure (photograph). Each exposure is to correspond to the photo number in the log. A brief description of the scene in the photograph is written down next to the photo log number. Also included in the photo log is a sketch of the scene to indicate the location of the photographer during each photograph. On the sketch, a circle is placed around the number indicating the photo log number with an arrow showing the direction the camera lens is facing.

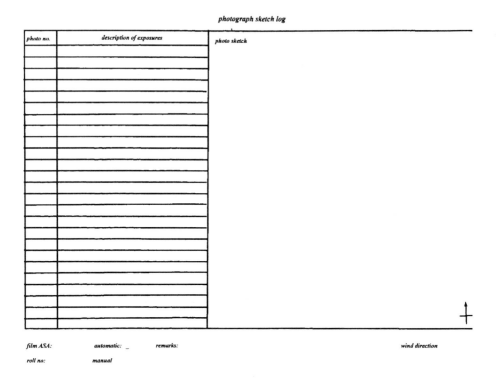

Figure 9-2. Example of a sketch photo log.

The first photograph will identify the fire scene through information gathered by the investigator and is used to help identify the film when the photographs are returned from the developer. This technique will eliminate the guesswork of deciding what roll of film goes to what fire scene. The information consists of the following: The investigator's name, the date the photographs were taken, the film roll number (on many scenes more than one roll of film will be taken and each roll of film will begin with this information), the location of where the photographs were taken, the agency involved, and if possible, the agency case number. This information is written with a thick black marker for clarity against a light color background. To aid the investigator, the information headings can be printed on the face sheet of the photo log.

photographer _____

date_____ **film roll number** _____

location_____

agency_____

agency case number _____

Figure 9-3. Information sheet of the photo log.

The second photograph should identify the actual fire scene through a street name, a numerical number, or a last name on the mail box.

Exterior Photographs

The next series of photographs should consist of the exterior exposures. Photograph each of the four elevations and each of the four corners beginning at the most prominent feature of the structure (usually the front) for identification, then, proceed in one direction (either clockwise or counter clockwise) and photograph all sides in a continual, systematic 360 degree circle. Photograph all windows and doors, any signs of forced entry, exterior fire damage, any type of evidence (footprints, containers, tools left behind, contents of the house left behind, etc.), outside sheds (if any), close-up of utilities (gas and electric along with their connectors, meters, fuse boxes, or any evidence of tampering) and any other item that may be of importance to the case.

Interior Photographs

After completion of the exterior shots, the next photograph will show the point of entry by the investigator into the structure's interior. In most cases, the investigator will begin photographing the area with least damage. Photograph all sides of the room, ceiling, and floor.

Photograph contents or lack of contents, furniture, and any type of fire, heat, smoke, and water damage. From this room, work towards the room of origin. Always look for evidence that indicates heat, fire travel, separate and noncommunicating fires, explosive and/or incendiary devices, tampered sprinkler systems, disconnected utilities, removable valuables, tampering of burglar or fire alarms, and any other items of importance. When fire damage is observed, photographs should illustrate heat and smoke patterns, char patterns, V patterns, melted metals and melted glass, floor patterns of low burning or liquid patterns (if present), and any other indicator that discloses fire behavior.

Once in the room of origin, photograph the area before disturbing any of the debris. Photograph anything unusual during the debris removal in layers. Once the point of origin has been established, photograph in detail the area of the heat source. If the heat source is electrical and involves an electrical appliance, try to get a close-up of the make and serial number. If the point of origin involves an incendiary fire, photograph each piece of evidence to document the crime.

The following photo log and photographs illustrate how to properly record the photographs and what the photograph should look like when developed.

The fire scene was a one-story CBS duplex lying in a east/west plane of direction, with the front facing north. Fire damage was contained to the east apartment. The apartment consisted of a small living room, one bedroom, a small kitchen, and a bath. This fire was used as a training exercise where two separate and noncommunicating points of origin were used.

Figure 9-4 is an example of what the photo log looks like when used in an actual fire scene investigation.

photograph sketch log

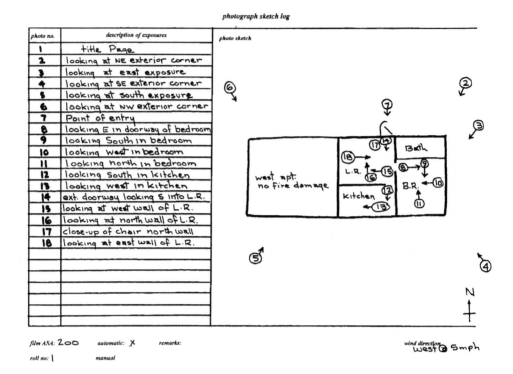

photo no.	description of exposures
1	title Page
2	looking at NE exterior corner
3	looking at east exposure
4	looking at SE exterior corner
5	looking at south exposure
6	looking at NW exterior corner
7	Point of entry
8	looking E in doorway of bedroom
9	looking South in bedroom
10	looking West in bedroom
11	looking north in bedroom
12	looking South in kitchen
13	looking west in kitchen
14	ext. doorway looking S into L.R.
15	looking at west wall of L.R.
16	looking at north wall of L.R.
17	close-up of chair north wall
18	looking at east wall of L.R.

film ASA: 200   automatic: X   remarks:

roll no: 1   manual

wind direction west @ 5 mph

Figure 9-4. A sketch photo log from an actual fire scene.

Figure 9-5 is a photograph of the information card. The photographs in Figure 9-6 depict the minimum number of photographs taken to document the exterior of the fire scene in a 360 degree circle.

Figure 9-5. Photograph of the information sheet (Photograph).

Figure 9-6a. Documenting the exterior of the fire scene with photographs (Photograph). North exposure, NE corner (photograph).

Figure 9-6b. East exposure, NE corner (photograph).

Figure 9-6c. East exposure, SE corner (photograph).

Figure 9-6d. South exposure (photograph).

Figure 9-6e. North exposure, NW corner (photograph).

Figure 9-7 is a photograph of the point of entry into the interior of the east apartment. Figures 9-8 and 9-9 are photographs that depict fire, heat, and smoke damage to the area of least damage, the bedroom and kitchen. Figure 9-10a is taken from the front exterior front door to reveal the back interior wall separating the living room and kitchen. Figures 9-10b, c, e document the remaining three walls of the living room. Figure 9-10b also shows the damage to the living room couch that was determined to be one of the points of origin. Figure 9-10d is a close-up of the north living room chair that was the second point of origin.

Figure 9-7. Point of entry into the interior (photograph).

Figure 9-8a. Damaged bedroom (photograph) East interior wall (photograph)

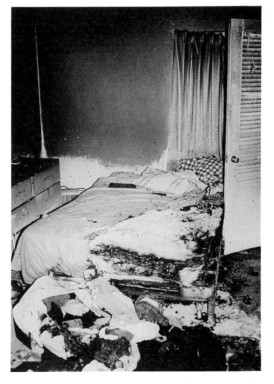

Figure 9-8b. South interior wall (photograph).

Figure 9-8c. West interior wall (photograph).

Figure 9-8d. North interior wall (photograph).

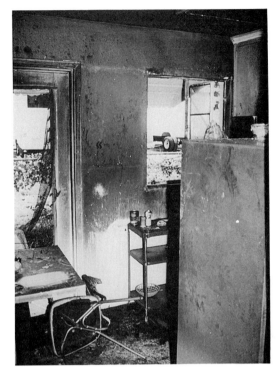

Figure 9-9a. Damaged Kitchen (photograph). South interior wall (photograph).

Figure 9-9b. West interior wall (photograph).

Figure 9-10a. Damaged living room (photograph).
The south interior wall taken from the point of entry (photograph).

Figure 9-10b. West interior wall (photograph).

Figure 9-10c. North interior wall (photograph).

Figure 9-10d. Close-up of the area of origin on the north interior wall (photograph).

Figure 9-10e. East interior wall (photograph).

The case file photo log is another type of log that can be utilized. When presented in this form, the organized report and photographs are an impressive presentation when submitted for review. Depending on the format of the photographs, a number of photographs can be fixed onto an 8 1/2 x 11 inch sheet of paper. Through the use of a copier machine, the investigator can construct the case file photo log adding space for important information. Figure 9-11 is an example of a case file photo log for 3 x 5 inch photographs.

date of fire_____   agency case number_____

case file photo log

	No. _____
	taken by:_____
	remarks:_____
	_____

	No._____
	taken by:_____
	remarks: _____
	_____

Figure 9-11. Example of a case file photo log.

PHOTOGRAPHING EVIDENCE

Tool marks leave a distinct mark and can be photographed as evidence. The first photograph is the area at large showing the relationship of the tool mark to the overall scene. The second photograph is a close-up of the mark. A third photograph may be made showing the mark even closer. The next photograph is a close-up of the tool mark with a scaled reference to show the size of the mark. When a tool is found at the scene, compare the end of the tool to the mark and photograph the comparison.

Always take two photographs involving evidence. First, a photograph is taken of the evidence exactly in the position as when it was discovered. Nothing is disturbed or removed until that photograph is taken. The second photograph involves the use of a scale reference to show the size of the evidence. A standard ruler, a tape measure, or a yard stick is placed alongside the evidence, then photographed. When documenting a pour or unusual burn pattern, photograph the pattern as found. To expose the pattern, mark the pattern's outline with a colored string, paint, or chalk and take a second photograph. The purpose of taking two photographs involves the admissibility of the photos as evidence. Using only one photograph containing extraneous props to help aid the interpretation of the photograph may be found inadmissible. The objection would be based on the fact that the photo may mislead or influence the trier of fact. Therefore, use one photograph in a natural, undisturbed position and the other with props.

If part of the determination of the fire cause was made through reconstruction, then photograph the reconstructed furniture, walls, floor, ceiling, and all sides of the room again.

PHOTOGRAPHING A FIRE FATALITY

In a fire fatality investigation, the first interior photographs should be views of the entire scene showing the location and position of the body. Overall views are to show the relationship of the body to the surrounding area (location of exits, placement of furniture, etc.) The photographs of the body are taken before any debris is removed. Once the overall views are taken, then close-up views are photographed. These include full-length shots of the body at all angles, photographs of different areas of the body (head, feet, hands, etc.); photograph any unusual wounds or injuries observed to the body. After this series of photographs is taken, remove the fire debris from the body and photograph the body again. The next series of photographs involves rolling the body over and photographing the underside of the body. Also, photograph the protected area of where the body was located, looking for weapons, clothing, indicators of how the fire was possibly started, and any other item that may be of value in the investigation. Once all photographs, notes, and sketches have been taken of the fatality, and the victim removed, proceed to document the fire scene. At the medical examiner's office, the body will be cleaned up as much as possible for the autopsy. The fire investigator or assistants from the medical examiner's office should then photograph the cleaned body.

AERIAL PHOTOGRAPHS

Aerial photographs are taken from any elevation above the fire scene. This can be accomplished through the use of aerial fire apparatus, law enforcement and fire department aircraft, another building, or a hill. If ground photographs of the fire scene are taken, the court may find aerial photographs inadmissible based on the photographs being cumulative. The purpose of taking aerial photographs is to show fire patterns and direction of fire spread. These observations cannot be observed and photographed on the ground. Based on this advantage, aerial photographs may become admissible if:
1. The property is properly identified.
2. An accurate portrayal of the fire scene is given in testimony by the photographer.

LIGHTING

Color is detected through different wavelengths of light. When all the wavelengths are observed, the color seen is white. When all the wavelengths of light are not seen, the color black is developed. The colors in between white and black have pigments that absorb a particular wavelength and reflect the others. With this in mind, the blackened interior from a fire will absorb the light. Almost every fire scene, therefore, will need some type of supplemental lighting to document the scene with photography. The source of light can come from an electronic flash, portable lights, special vehicles designed for lighting, and in an emergency even the flashlight.

Electronic flash equipment is made to resemble daylight when energized. It is a strong light that is directional and enables the photographer to take photographs in poor lighting. By consulting manufacturing handbooks and reading articles and books, the photographer can experiment on the aspects of flash photography.

One experiment an investigator can practice is called "Painting with light" and is used at night to illuminate an area. This technique uses a camera located in a stationary position with the flash unit moving to different positions to illuminate the object. The following equipment is needed to successfully complete the experiment: one electronic flash (or another light source), a camera that must have T or B setting (a T or B setting lets the photographer open and close the shutter on demand), a tripod and a locking shutter cable release. The camera must be on a tripod and placed towards the subject at eye level. Looking at the subject, make sure strong lighting cannot reflect back into the camera's lens (street lights, vehicle lights, etc.). Looking into the viewfinder, determine the viewing area both to the left and to the right by using the beam of a flashlight on the structure. Decide how many flashes will be necessary to paint the structure. This can be determined by knowing how many feet across the flash unit will light up the subject. Set the lens opening to f 3.8 or larger. The shutter speed is set on either T or B. Focus at the half-way point between the camera and the furthest point to be photographed. Use only one frame of the film, DO NOT ADVANCE THE FILM. When the preplanning is completed and the photograph is ready to be taken, two people are used. One individual is placed by the camera on the tripod with the shutter properly set. The second individual works the flash unit at the desired locations and must be behind the camera's view. When everything is in place, verbal communication is important. The individual working the flash unit states, "ready." The camera person releases the shutter and states "go." The flash unit is then energized. After the flash, the person working the camera immediately carefully covers the lens with a dark surface to stop unnecessary light from entering the film. At the new flash location, the communication is repeated. This process is repeated until the desired effect is completed. Once the prearranged number of flashes has been activated, release the shutter cable. If other people are present, make sure they are away from and behind the camera to avoid jiggling the tripod. Flashing behind the structure will give a sharp, distinct outline of the structure.

The following series of sketches and photographs of a darkened building demonstrate the use of painting with light The sketches on the left show the relationship between the camera, the flash unit, and the object. The photographs next to the sketch exhibits the effect of the flash sequence.

Figure 9-15b shows what can happen when the individual working the flash unit is in front of the camera's viewing. The lower right corner of the photograph reveals a silhouette of the flash unit operator.

Figure 9-12. Painting with light using one flash (diagram and photograph).

Figure 9-13. Painting with light using two flashes (diagram and photograph).

Figure 9-14. Painting with light using three flashes (diagram and photograph).

Figure 9-15. Painting with light with a person in front of the camera lens
(diagram and photograph).

Figure 9-16. Painting with light on a vehicle using one flash (diagram and photograph).

Figure 9-17. Painting with light on a vehicle using two flashes (diagram and photograph).

Figure 9-18. Painting with light on a vehicle using three flashes (diagram and photograph).

Painting with light can also be used outdoors for automobiles, boats, airplanes, or any other exterior crime scene. Through preparation and decisions as to what is needed in the photograph, painting with light can expand the illumination of a darkened scene to reveal evidence left on the surrounding ground.

The following series of photographs depict a vehicle on a dirt road. Three different flashes were used to light up the scene. More flashes could have been used if the background was to be lit.

In an explosion atmosphere or when the flash unit fails for whatever reason, a flashlight (a sealed flashlight when used in an explosive atmosphere) can be used to paint the scene with lights. The identical procedure used for setting the camera with the flash unit is performed. When the shutter is opened, use the flashlight like a paint brush and slowly go up and down and back and forth on the object. Do this several times. When painting with the light, overlap the areas and extend the area two feet out of the camera's range. Areas closer to the camera can be given less light. For best results, bracket several exposures using this technique to ensure a correct exposure.

Painting with light ensures that important features of a fire scene investigation will not be lost due to darkness, or the failure of lighting equipment.

To help the investigator use lighting on the fire scene, here are some helpful hints:

- When shooting with flash, bracket the image to help get the correct exposure. This is easily done by finding the appropriate f-stop (size of the opening in the shutter when released). The photograph is then taken at this f-stop and the next f-stop below and above for a total of three exposures of the same image. (This cannot not be performed with point and shoot cameras.)
- When photographing the depth of char in alligatoring of wood, burn patterns, and spalling, use what is called "cross lighting."
  This procedure requires the light source to be at an angle to eliminate reflections. The same procedure applies to window glass. Photographing directly at the glass will cause the flash to bounce back at the camera and cause what is called "whiteover."
- To show the different types of heat stress on window glass, use a light source on the outside as a backlight. The photograph is taken from inside the room.
- Fill flash can help in daylight photography by lighting up deep

shadows that would form through natural lighting. These shadows may hide important evidence from view.
- When using a flash make sure the atmosphere is explosive free. Remember the electronic flash generates a considerable amount of heat.

MAINTENANCE OF THE CAMERA

A fire investigator's camera operates in a harsh environment. It makes sense then to care for your camera. A clean and well-maintained camera can add to the camera's life. By following a few basic rules you can get many years of service from your camera.
- Maintenance begins by learning to properly operate the camera through the manufacturer's manuals.
- On the fire scene, protect the camera whenever possible from heavy bumping or dropping. Never carry the camera by the strap. This will cause the camera to rotate and swing around when you walk.
- When the camera and its accessories are not in use, they should be enclosed in a protective case constructed of a hard material.
- Significant changes in temperature will cause condensation on the lens and the camera's body.
- Protect the lens with a haze filter when shooting and a lens cap when not shooting.
- After using the camera in a fire scene, remove dirt, grease, and moisture with a soft cloth, using a manual air brush. If smudges are found in the viewfinder or the glass of the lenses then a small amount of lens cleaner, not alcohol, may be used on a cotton swab and applied lightly on the glass.
- Dry away any area exposed to water.
- Do not use lubrication on the camera, have all servicing done at an authorized repair service, unless, of course, you are familiar with camera repair.
- If the camera is not going to be used for several weeks, then remove the batteries to prevent battery acid drainage.
- When in a humid climate, prevent molds and fungus by storing camera and accessories in an airtight container with a package of dry silica.

- Older cameras of the mechanical type need cleaning, lubrication, and adjustment more often than the electronic models.
- Cameras in heavy use and in harsh environments should be serviced every six to twelve months.

Many law enforcement and fire department photographers have compiled a list of tips of things to do and not to do on a fire scene. The following tips will start you on your way to developing good habits while photographing:

- You want the fire scene to be a true and accurate representation of what you saw; therefore, do not photograph cigarette packs, discarded cigarettes, styrofoam cups, or film packaging from the personnel involved in the investigation. KEEP A CLEAN SCENE.
- When shooting floor patterns, do not show the tips of your feet or the feet of other investigators.
- Do not photograph individuals in scene shots unless there is a specific purpose to do so.
- Do not write on the back of photographs. Use stick-em paper to write on.
- What should be photographed in a fire scene? Everything! Outside the fire scene, all exposures openings and evidence. Inside each room to show damages, contents, furniture, and evidence.
- The amount of photographs taken is up to the discretion of the investigator. As a general rule, it is better to take too many than to take too few.
- Still photography should not be replaced by video. The use of video is to be used in conjunction with the stills.
- Label each film canister after the roll is completed to avoid confusion as to what roll is what.
- Do not add another fire scene on the same roll.
- Carry extra camera and flash unit batteries.
- When the effect of a wide angle lens is needed, but is not available, then overlap the exposures or frames. Pick a reference point on one side of the image and start the next frame from that point. By combining all exposures, a panoramic view called "mosaic" will be developed. For best results use a tripod.

ADMISSIBILITY OF PHOTOGRAPHS

Photographs, as demonstrative evidence, help the witness present evidence through testimony. Words alone may be difficult for the trier of fact to understand. The expressed words of the testimony, combined with photographs, bring together an understanding for the trier of fact. For example, a fire investigator, trying to describe V patterns in a fire scene without photographs must do so through the descriptive word.

The trier of fact hears the words and forms a mental picture of what is said. Now, take the same descriptive words and at the same time point to a photograph depicting a V pattern. There is no doubt that the trier of fact will now understand what a V pattern represents. As a general rule, without the witness, photographs cannot stand alone as evidence.

One of the best descriptions in a court case to illustrate why photographs are used in court comes from the case, *Barham v Norell* (138 So. 2d 493, 243 Miss. 441) in 1962. The court stated:

> The accuracy of verbal representation of a scene is subject to many limitations. The person who makes the observation is not likely to observe with complete accuracy. Language as a means of communicating the representation to another has its limitations. Human memory is faulty. And scenes verbally represented can be consciously or unconsciously misrepresented. Pictures can also be distorted, but not in the same way and not as easily as verbal representation.

Therefore, a photograph to become admissible must be relevant to the case and verified by the witness. To be relevant, a photograph must meet at least one of the criteria listed below:

1. The photographic evidence must have a reasonable tendency to establish or disprove issues in a case. The issue is not really the photograph, but the subject matter reproduced in the photograph.
2. If the photograph represents evidence that the trier of fact could have examined and viewed at the time the image was taken.
3. When the photograph assists the trier of fact in understanding the case. The photograph should be an instructive visual aid. The photograph cannot confuse or mislead the jury. The photographed evidence is used to corroborate the oral testimony or other evidence in both a civic or criminal case.

4. When the photograph tends to disprove, contradict or impeach a witness. To disprove a witness, the photograph should be shown so that the witness can explain or contradict the photograph.

Photographs, then, are used to assist a witness in illustrating or explaining his or her testimony. These photographs cannot show unfair prejudices, confuse the issues, mislead the jury, cause undue delay, waste time, and be a needless presentation of cumulative evidence.

Since photographs can be altered by man, verification by a trustworthy witness is required. The witness must testify that the image in the picture is a fair representation of what they observed. Based on this fact, the incident should be photographed during the event. Since this seldom happens the scene should be photographed as soon as possible after the incident. The witness must vouch for the photograph's truthfulness and accurately identify the person, place, and location of evidence in the picture. It is preferred that the one who took the photograph be the witness. However, any other witness who was at the scene, can attest that the picture is a fair representation and can state what the photograph depicts. The witness must be able to identify a fingerprint, palm print, footprint, tool mark, or any other piece of evidence found at the scene as the same image observed in the photograph. After photographing the evidence in its natural state, some type of identifying mark can be made. The witness' initials can be scratched into the surface of a tool, ID tags with the date, time, name of witness, and location where taken, attached to the evidence.

In the Supreme Court case, *Wilson v. United States*, (162 US 613, 40 L ed 1090) the court stated to have a photograph admissible in court, the photograph must be:
1. Properly taken.
2. An accurate likeness.
3. An accurate representation of the subject in question.

The admissibility or rejection of photographed evidence is up to the discretion of the trial judge. The courts look favorably on photographs because they can clarify issues and give a clearer understanding of the physical facts.

There are two basic theories on photographed evidence being accepted in the courts today. The first theory is called "Pictorial Testimony." This theory involves testimony of a witness who uses photographs to help illustrate his or her verbal communication. The second theory is called "Silent Witness" which is a photograph that speaks for itself. A Silent Witness photograph can speak in court silently by not using speech but images of persons, places, and things by a camera that is time tripped or automatically left on. The photograph does not illustrate the testimony of a witness but is the witness to something happening. An example of Silent Witness photograph is an X-ray where a witness can clearly testify from inside the body without actually being there, simply by looking at the photograph. A second example involves a still camera used as a security precaution that is activated when a perpetrator illegally enters a building and is photographed setting fire to the structure. The camera is the only witness to the crime. A witness, who in this case does not have to be the photographer, must testify as to when, where, and under what circumstances the photograph was taken, plus the accurate portrayal of the subject(s) in question.

The defense may object to having a photograph admitted into evidence on many points. Some of the most popular objections are:

- Accuracy of color–does the photograph represent the true color of the scene?
- Was the negative or print altered in any way?
- Enlargement over 8' x 10' makes the image larger than life and is not a fair and accurate representation. It may be necessary to explain the purpose of the enlargement.
- Errors in printing due to negative reversal?
- Types of equipment at one time was a factor while testifying in court. The photographer had to testify as to how the photograph was taken and by what means. Today, with the electronic age and automation, the photographer just points and shoots. The defense attorney may try these old tactics.
- Inflammatory photographs that are graphic in detail and may be prejudicial to the jury. If the inflammatory photographs are relevant and verified, they may be admitted.
- Optical illusions that are created by using lenses that distort vertical and horizontal lines. Such a lens is the wide angle that can give appearance of looking into a fish bowl.
- When the time frame of when the photograph was taken is further away from when the incident occurred. Due to changes in appearance between point A, the incident and point B, the photographs taken of point A may not show a fair and accurate representation.

- When the photographs offered were not relevant to the facts correctly.
- When photographs are cumulative, that is a set of photographs have already been submitted depicting the same images. It is best to review all the photographs in the case and use the best and most effective photographs first. Photography, then, can be used as a tool to record the fire scene, document evidence and establish origin and cause. In the courtroom, photographs support and verify the testimony of the witness.

Photography, then, can be used as a tool to record the fire scene, document evidence, and establish origin and cause. In the courtroom, photographs support and verify the testimony of the witness.

Chapter 10

VIDEO

Like the still camera, the video should be another tool used to document the fire scene. If the video is taken properly, it can travel over the scene like the human eye and observe the position and relationship of objects, just as the investigator observed. Video is not limited to just a moment of time, as in still photography. Instead, videotape technology captures an image showing action or, sound and action, adding the dimension of a scaled time that can be shown in sequence. The primary use of the video is to describe an event as it occurs. In a fire investigation, a video is a moving documentation of what a fire scene looked like from start to finish. Since the video is a reproduction, it has the capability of bringing the complete fire scene into the courtroom as an aid to the witness and to help the trier of fact understand the issues in the case. The video can also be played over and over again to help the investigator in report wiring and to refresh the memory of a witness during interviews. It is important to remember that video is used in conjunction with still photographs, sketches, and field notes, and IS NOT used as a substitute for them. Combining these forms of demonstrative evidence will document the fire scene and visually prove the event took place.

On the fire scene, the investigator should understand that the video production is not for entertainment, nor to experiment with special effects. The purpose of the video, like any other type of demonstrative evidence, is to depict a fair and accurate representation of the scene.

HISTORY OF VIDEO

The history of video began, not in the electronic age, but during the early stages of modern photographs, called motion pictures. Since the invention of still photography, man had been trying to capture the sense of movement on film. One of the first individuals to establish himself in motion pictures was Eadweard Muybridge, between 1872 and 1877. Muybridge, in California, captured the movement of a running horse by having the horse trip several wires in its path that were connected to still cameras. The photographs were then mounted on a stroboscopic disk and projected by a lantern. The projected image reproduced the running action of the horse.

During one of Muybridge's lectures on motion pictures, another individual by the name of Thomas Edison became curious in this new concept. Through assorted experiments Edison patented a motion picture camera in 1887. However, the problem with the camera was that an image could not be reproduced. An assistant of Edison, William Kennedy Laurie Dickson, used the celluloid film developed by George Eastman and Edison's camera to produce several 15-minute films. However, the credit of inventing motion pictures goes to Thomas Edison with his camera and an electrically-driven peephole viewing machine that he developed called the Kinetoscope. From this point on, motion pictures rapidly progressed both in Europe and the United States.

Movie cameras captured on a roll of film the action of images and the dimension of time onto thousands of small, two-dimensional still photographs. After development, the roll of still photographs are shown through a projector at a fast speed, giving the illusion of movement. Depending on the type of camera used, sound may be recorded as well. Since the roll of movie film could have been edited or distorted easily, the courts made it very difficult for the motion pictures to be admitted into evidence. Only through extensive testimony stating that the film was properly processed, properly taken, and properly projected was it then allowed to be admitted. Today, the reliability and accuracy of the film is not based on its creation, but on the testimony of a witness that the reproduced scene is how he or she observed the scene or events.

The first court case, *Duncan v. Kieger* (6 Ohio App. 57, 1916), involved motion pictures as evidence in 1915. The evidence was used in a personal injury suit when the plaintiff filmed the defendant walking along a street better than he claimed he could in court.

Sound motion pictures were held admissible in 1930 in the case *Commonwealth v. Roller* (100 P. Super 125). The court allowed the photographed confession into evidence on the basis that the law is receptive to any reliable mechanism that can produce evidence.

The video age began in 1956 when the first video tape recorder was introduced into the television world at a CBS affiliates meeting in Chicago. These video recorders were reel to reel and produced only black and white pictures.

During the late sixties, police agencies across the United States started using videotapes as evidence in police line-ups, video shots of arrested suspects, confessions, reenactments of crime scenes, and the documentation of crime scenes. As the value of the video became apparent, fire departments saw a need for the video in training, public education, and fire scene investigation.

In 1970, Sony introduced the U-matic video cassette player, better known today as the camcorder. The word camcorder comes from two words—video camera and videocassette recorder. The camcorder has become the standard format used today because of its size, easy use, and lightweight camera.

The difference between videos and motion pictures is in the way the images are recorded. As stated earlier, motion pictures are produced on film. The video, however, records images and sounds into electronic impulses on magnetic tape. In this form, pictures cannot be visible until played back.

The following sections will discuss the types of cameras, how to use lighting, sound, composition, and special effects in conjunction with the camera. The last section will discuss the video camera and the legal system.

TYPES OF VIDEO CAMERA

Before knowing how to use the video camera, the choice of what type of camera to use should be made. The larger format video cameras are good for the professional video user, but not for the fire investigator. Known as the 3/4-inch format, these cameras are bulky and heavy to use, especially in the tight quarters of interior fire scenes. The 3/4 camera is a sensitive piece of equipment that cannot withstand the constant conditions that exist in the fire scene environment.

Since the late 1980s, the camcorder video camera has become the camera of choice for both the amateur home video maker and the crime scene technician. The camcorder combines both the camera and the recorder into one small unit. The basic formats used in the camcorder are the VHS and the 8mm.

The VHS (Video Home System) is a full-size camcorder with a 1/2-inch tape. Depending on the length of the tape, the camcorder can record for a maximum of two hours and forty minutes. A typical layout of a VHS full-size camcorder and the camera's basic features are shown in Figure 10-1. A variation on the location and number of features will depend on the make and model of the camera. Most VHS-C, 8 mm super VHS, Hi8, and 3/4 format will have similar features. Before purchasing a camera, consult the owner's manual to verify that the camera contains the desired features.

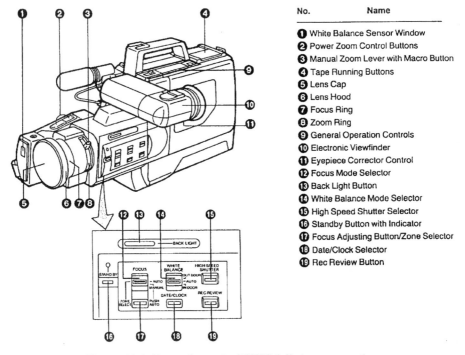

No.	Name
❶	White Balance Sensor Window
❷	Power Zoom Control Buttons
❸	Manual Zoom Lever with Macro Button
❹	Tape Running Buttons
❺	Lens Cap
❻	Lens Hood
❼	Focus Ring
❽	Zoom Ring
❾	General Operation Controls
❿	Electronic Viewfinder
⓫	Eyepiece Corrector Control
⓬	Focus Mode Selector
⓭	Back Light Button
⓮	White Balance Mode Selector
⓯	High Speed Shutter Selector
⓰	Standby Button with Indicator
⓱	Focus Adjusting Button/Zone Selector
⓲	Date/Clock Selector
⓳	Rec Review Button

Figure 10-1. Parts of a typical VHS full size camcorder.

With the trend to have items smaller and lightweight, the makers of the VHS camcorder have followed in the same direction and developed the VHS-C. The C stands for compact because of the smaller 1/2-inch size tape cassette and camera. The recording time of the tape in standard play is approximately 20 minutes with extended play recording for an hour.

To help reduce the size and weight of the camcorder, a smaller format was developed called the 8mm. This type of camera has the same picture quality. The standard speed of the tape will record for approximately two hours. The composition of the tape is different than the 1/2 inch. The 8 mm tape contains small metal pieces for a higher signal output. The small size can become a disadvantage when trying to keep the camera steady and comfortable. The same basic features found on the VHS can be found on the 8 mm.

The latest technology has developed a Super VHS and a Hi8 with advances in both the tape composition and the camcorder and VCR deck. This technology has improved the picture quality, contrast, and color.

The format best suited for the fire scene investigator is the VHS or the 8 mm. Since the investigator is documenting the fire scene, the camcorder should be a simple and straightforward piece of equipment with the basic features needed to produce the desired fire scene investigation video.

LIGHTING

With the darkened remains of the interior of the fire scene and the lack of natural and artificial light, lighting becomes an important element in a fire scene video. The camcorder of today is designed to use the average light available in the average home. This light is measured in a unit which measures the brightness of light called lux. For example, the brightness of a clear sunlit midday sky has the measure of approximately 100,000 lux. The brightness of a candle flame 12 inches away is measured at 10 lux. The lux rating of the camcorder is determined by a numerical value. The lower the number, the more sensitive the camera is to light.

To understand how to use lighting on a fire scene, basic knowledge of lighting terminology and techniques will be examined.

To have optimum lighting on any subject, the three-point lighting setup, involving the key light, fill light and the back light is often used as illustrated in Figure 10-2. Before using the three-point lighting setup, the subject should have enough base light to illuminate the subject so the other lights may be used. The key light is a hard light used as the main lighting source on the subject. The key light is located slightly above the camcorder pointing down at an angle of 10 to 40 degrees and 10 to 30 degrees off to the side of the camcorder. The fill light is used to fill in any shadows that may be caused by the key light. The fill is a soft light located anywhere from 10 to 30 degrees on the opposite side of the camcorder and at the same height as the camera. Whenever possible, have the fill light as a broad flood light instead of a spotlight. If deep shadows still exist after applying the fill light, then readjust the angle of the key light. Also, raising and lowering the key light can help eliminate shadows. The back light is used behind the subject to separate it from the background. The back light is above the subject and pointed down at a steep angle. The three-point lighting setup is excellent to use when the subject is stationary, such as in an interview or an interrogation.

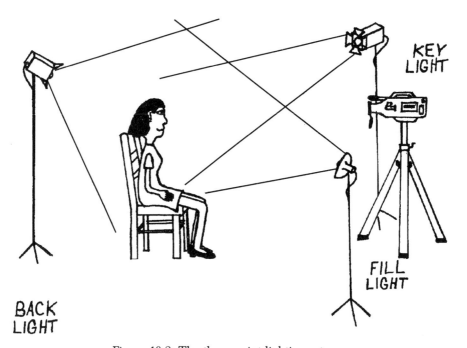

Figure 10-2. The three-point lighting set-up.

On the fire scene, especially with interior shots, the three-point lighting setup is not practical to use. Most video shots of the interior would be too time consuming to set up and work around the fire debris. The minimal amount of artificial lighting to use on the fire scene can be the single video lamp. This type of lighting, however, will create a flat effect with shadows casting in the background. Also, a direct light beam hitting the subject being filmed may cause the color to wash out. To solve this problem, an adaptation of the three- point lighting may be utilized. This is called the two-point lighting setup and is illustrated in Figure 10-3. The two-point system can give the image texture, color saturation, and dimension. To achieve the two-point setup, use the base light as the fill light and the video light as the key light. The fill light should be at least one high wattage portable halogen scene light at 10-40 degree angle from the camcorder to flood the complete area to be filmed. Again, be careful of washout from too much lighting. The attached video camcorder light then becomes the key light as the light illuminates whatever is in front of the lens. If the situation arises that the only source of light is the video camcorder light, then use it directly in front of the subject. The light should be soft and in a flood mode. If the light is adjustable, set it between spot and flood.

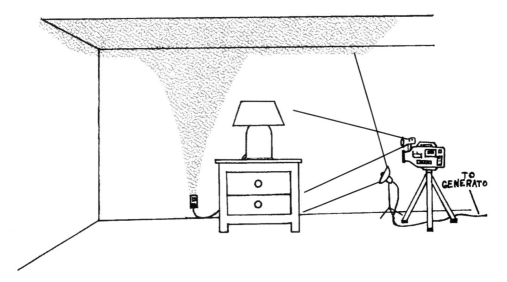

Figure 10-3. The two-point lighting for fire scenes.

When shooting during the day, the operator of the video camera must always be conscience of the ambient light. If sunlight is used, it can be so strong that the camera will respond to it by causing the subjects to become silhouettes. This condition will happen almost every time an interior scene is shot and strong daylight comes through windows or an exterior opening. What occurs is the camcorder's exposure system closes down due to the strong back lighting making the subject appear as a silhouette. To overcome this problem use a light source.

SOUND

The sound on the video is picked up by the built-in microphone on the camcorder. It is recommended that the camcorder be equipped with an external jack to help enhance the sound quality by using a better quality microphone.

There are two types of microphones, unidirectional and omnidirectional (shown in Figure 10-4) that can be used. The unidirectional microphone picks up sound in a pattern of 80 percent directly in front of and 20 percent from the sides and back of the microphone. This sound pattern is good for most general video recordings. When using a unidirectional microphone, remember that the closer the microphone is to the subject being filmed the better the sound quality.

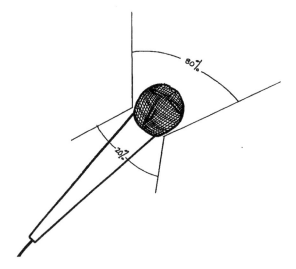

Figure 10-4a. The unidirectional microphone.

The omnidirectional microphone is not selective and will pick up sound equally in all directions. The omnidirectional microphone does not work well on a fire scene narration or during interviews due to the problem of picking up unwanted background. Based on this 360 degree pickup, the camera operator must be quiet. Sometimes even the breathing of the camera operator can be heard. The camera operator should also be aware of any conversation taking place behind the camcorder.

Figure 10-4b. The omnidirectional microphone.

"QUIET ON THE SET" is the phrase often heard on television and movie sets before cameras begin filming. This rule should also apply on the set of the fire scene. Once the camcorder starts recording, it must be remembered that not only is the visual being recorded but also all ambient sound. It can be very embarrassing to the investigator when he/she presents the video of a fire scene at trial and hears in the background two individuals telling jokes to one another. If sound is not important to the video and the camcorder has an external audio jack, then use an unattached plug to shunt the built in microphone.

The audio of an interview or interrogation must be clear and comprehensible. Each individual involved in the interview should have a unidirectional microphone in front of him or her. Each microphone is plugged into a sound mixer that regulates the volume and generates the sound through one electronic cable into the camcorder. The sound mixer is easy to use, not very expensive, and can be purchased at most electronic stores.

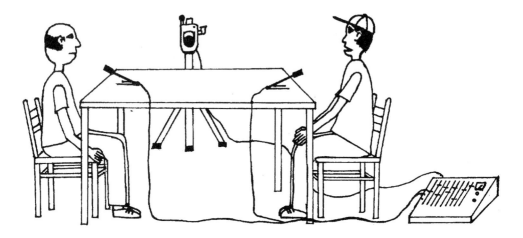

Figure 10-5. The placement of microphones and use of a sound mixer during interviews.

Another inexpensive piece of sound equipment that can be utilized is the FM wireless microphone. With the tether of the audio microphone cord to the camcorder absent, the fire investigator can narrate the fire scene without the cord being pulled through the fire debris. The investigator can now concentrate on the path of travel and the narration. The FM wireless microphone can also be used on the interview, doing away with the numerous microphone cords.

Whenever a videotape is to be presented at trial, a typed transcription of any dialogue should be furnished to the trier of fact. This written record can assist in following any inaudible dialogue that may occur due to a malfunction of the audio segment of the tape or noise from the courtroom. Being well prepared strengthens the investigator's testimony and with a little effort in making the transcript, the investigator will encourage a favorable rapport between the investigator and the trier of fact.

COMPOSITION

The use of the television screen is a powerful tool that can be used to communicate the fire scene to the trier of fact. The investigator must control the action and not let the action control the investigator. It takes practice to develop a fire scene video. One cannot take the camera and take random shots or sequences that simply roam. There must be a reason for every shot taken. Practice will help the investigator start thinking of how the camera sees the subject. Look at the subject and think of how you can best project the image. What portions of the images are needless and can be eliminated from the frame. What is the story the video is telling? The answers to these and other elementary questions involving the fire scene video can be found in what is called the composition of the video.

The composition of the home video differs from the composition of the fire scene video. While the purpose of the home video is entertainment and amusement with a story that is told through different sequences with special effects and assorted sounds, the purpose of the fire scene video is to document facts and events as demonstrative evidence. The video is to be used as an aid to the investigator and is not a form of entertainment that is found in the home video. The production must be as elementary as possible using special effects to a minimum.

The composition of the fire scene video begins with the story that should be told in the simplest form with a beginning, a middle, and a conclusion. The story is broken down into a number of sequences. Each sequence is a collection of shots that are connected, centering on a specific action in a specific location. The shot is the basic unit of the sequence. The three primary shots used in a sequence are the long shot, the middle shot, and the close-up shot. The long shot (Figure 10-5) establishes the sequence by letting the viewers orient themselves to the setting. On the fire scene of a structure, the long shot would be a complete exterior view of the building. An interior long shot of a fire damaged room should show, as much as possible, the complete area. The middle shot (Figure 10-6) brings the image in closer but still leaves enough of a background so the viewer knows where the scene is being shot. The middle shot of the exterior of a fire damaged structure would contain the windows, doors, and any other physical evidence important to the investigation. The middle shot is actually used to sep-arate the long shot from the close-up shot and establishes the close up which holds the greatest impact of the sequence. The close-up shot (Figure 10-7) dominates the image of interest in the picture frame and grabs the attention of the viewer. The close-up shot of the exterior of a fire damaged structure could be the forced entry marks on the front door, the position of the windows, or the identity of any other pieces of physical evidence. By using the long shot, the middle shot, and the close-up shot, the trier of fact will get a clearer understanding of what the fire scene portrays. Imagine showing the trier of fact a close-up shot of an incendiary device without the long or middle shot. This is a common mistake made by investigators when they present their video. By deleting the long and middle shots, the trier of fact would not know where the device was located in relation to the overall scene. This type of practice can lead to inadmissible video as evidence dur-ing your testimony in court.

Figure 10-6. The long shot (photograph).

Figure 10-7. The middle shot (photograph)

Figure 10-8. The close-up shot (photograph).

Another mistake made by the amateur video user is the placing of the subject of interest right in the middle of the frame. Professionally speaking, "dead center," as this is called, is very boring and does not generate tension. When looking through the viewfinder of the camera, mentally divide the screen into thirds both vertically and horizontally. Place the subject of interest within one of the intersecting lines that will be acceptable to the eye.

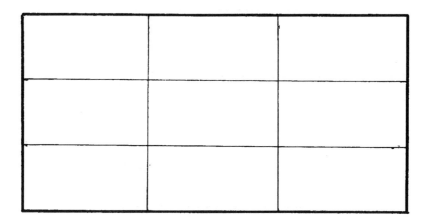

Figure 10-9. Dividing the view finder into thirds.

For example, in Figure 10-10, the action in this scene involves a basketball player dribbling the ball down the court. By placing the player to the left third of the screen, the viewer feels the motion to the right. The large space to the right also gives the viewer a clear area for subjects or objects to enter the picture naturally.

Figure 10-10. Placement of a basketball player in the view finder.

When using the video for an interview either the close-up shot or the middle shot will be the dominate shot of the interview. If the camera operator is experienced a combination of the two shots can be used to add dimension to the video.

In the interview, the focal point of the close-up shot will be interviewee's face. Anyone watching the video interview will focus their attention to the facial expressions surrounding the eyes. For this type of effect, the eyes are placed a third of the way down from the top of the screen with the subject's head just off center in either direction. This type of view can be seen in Figure 10-11. Using the close-up shot makes the subject the center of attention with minimal background distractions. Also, from this view, the subject can be brought into a middle shot to observe non-verbal body movement when asked specific questions.

Figure 10-11. Placement of subject in close-up shot during an interview.

Figure 10-12 is another form of attracting the facial expressions of the viewer. The technique uses a close-up shot of the subject, with the focal point on the left or right vertical third of the screen with the eyes, again, on the horizontal upper third. This type of shot is used extensively by the television industry in news and documentary footage. A word of caution: before filming, make sure the background does not take away the attention of the viewer from the subject. This method can also be used on inanimate objects to enhance the emotion of the viewer.

Figure 10-12. Placement of subject off center during an interview.

"Dead center" is not a rule that must be adhered to. There will be times when the camera operator will want the subject to be the center of attention. In Figure 10-13 the camera operator felt a strong emotional relationship between the father and daughter. To emphasize this emotion, a close-up shot was used and the father and daughter were placed "dead center" in the viewfinder.

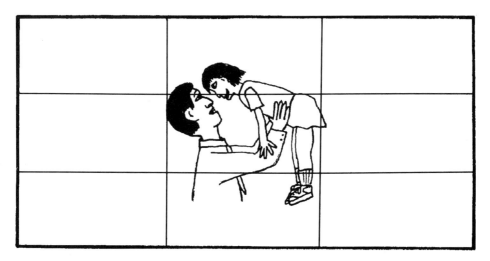

Figure 10-13. Dead center.

It is up to the investigator's creativity as to how the image will be perceived. As a composition tool, the frame of the image can be exciting and pleasing to the viewer.

SPECIAL EFFECTS
Position of the Camera

The camera's position and angle can influence how the image is perceived. The camera angle can dictate the audience position when viewing the image. It is best to shoot the fire scene at eye level. In this way, when the trier of fact views the video, the image will appear to be normal, a true and natural representation of how the investigator observed the scene. Any other angle or position of the camera may give an exaggerated and dramatic view. For example, in Figure 10-14, the camera angle is higher than the subject and is called a high angle shot. The image will appear smaller than its actual size, giving the viewer a sense of superiority and domination. In Figure 10-15 the camera is located lower than the subject and looking up. Called a low angle shot, this technique will magnify the image. The viewer has a sense of submission or helplessness. The investigator should always have a purpose and a reason if a particular angle is used. Remember that any type of distortion may cause the video to be inadmissible.

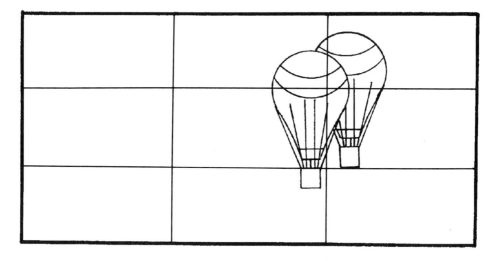

Figure 10-14. High angle shot.

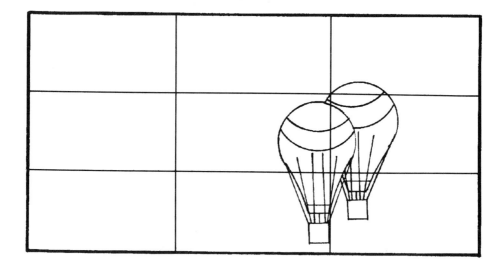

Figure 10-15. Low angle shot.

Zoom Lens

One of the few special effects that can be used, with a degree of caution, is the zoom lens. All camcorders have the zoom lens, allowing the view to be either an all encompassing wide angle view, up to a close-up or telephoto shot. The different views can all be conducted while standing in one position.

The problem with using the close-up shot of the zoom lens is distortion. At the edges of the frame, straight lines can bow out and vertical straight lies can bow in. This distortion is similar to the wide angle lens used in still photography. The courts may see this distortion as unrealistic and determine this video sequence inadmissible.

Another problem that the amateur camera operator may create when using the zoom lens is the quick zoom-in and quick zoom-out shot. This technique can give the viewer a choppy, almost seasick motion. If the zoom is to be used, several considerations should be made. First, the purpose of using the zoom-in feature is to underline a point or issue. Therefore, determine before shooting if the subject needs to be accentuated with the zoom. If the decision is made to use the zoom, stay with the following procedure for excellent results. Before activating the video recorder, zoom in on the subject as close as possible, then focus. This will allow the subject, during the filming, to stay focused throughout the zoom out, or, to the position desired for the beginning shot. Second, it is better to zoom out of a close-up shot than to zoom in. Without practice, zooming in can cause the subject to become shaky and out of composition. As demonstrative evidence, zooming in too close may exaggerate the subject and be taken as a means of causing a dramatic effect. This effect may cause the video to become inadmissible. Third, never shoot an object just in the close-up shot. The trier of fact, as a viewer, will have no idea where the object was located in relation to the scene.

As a rule of thumb stay away from using the zoom lens. If a close-up is needed, stay in the wide angle and either move in or out by walking.

Panning

Another form of special effects, called panning, is the horizontal movement of the camera from left to right or right to left from a fixed position that follows an action or to make some point of emphasis. Some video camera operators prefer the left-to-right movement because of the natural motion of the eyes when reading—left to right. Whatever direction chosen, a starting and stopping place should be established. The pan is to be a slow movement so that details are not overlooked and a dizzying effect is not produced. For best results, a preplan of the area of the pan can determine how much time is needed for the viewer to comprehend the subject matter.

The pan can be accomplished either by using a tripod or by holding the video unit. When a tripod is not available, execute a pan by using the camera operator's body as tripod. To do this, the camera operator should plant his or her feet about two feet, to two and one-half feet apart (think of the legs as a two legged tripod). Hold the camcorder on the shoulder. Turn the feet in the direction where the pan is going to stop. With a conscience knowledge of balance, turn the body in the direction the pan is going to start. Begin filming and slowly turn to the point where the camcorder is facing the same direction as the feet. For fire scene video, the pan comes in handy where a large area is being filmed and when filming interior shots of rooms. In essence what will be completed is a panoramic view of the scene.

Date/time Indicator

Most all camcorders have a date and time indicator that can be superimposed onto the recording tape. This camera feature allows the date and time to appear in a small area of video and verifies that the fire scene was recorded during the investigation. The verification is an important rule of admissibility when used during the trial. Based on this fact, it is recommended that the date and time generator be left on throughout the recording of the fire scene. When the camcorder is turned on, then off, then on again, the time clock will reflect the time interval that the camera was not recording. This time duration should be noted by the camera operator and under what circumstances the interval was made. By using this procedure, if questioned, the camera operator can explain through documentation how and why the camcorder was used. For example, the video begins with an exterior shot of the fire scene with the date 2-5-92 and the time of 12:30 PM. The camcorder is then turned off so that preparation for an interior video can be made. The date will remain the same and the time is 12:34 PM.

Once inside, the camera person begins the video of the interior, the date 2-5-92 and time 12:39 PM will appear. There is a time interval of 5 minutes the camcorder was off. The camera operator will document that time as 12:34-12:39 PM—preparing for interior shot.

Remember, creativity is not a fundamental interest of the camera operator on the fire scene. What is of interest is how the video will portray a correct and accurate representation of what the investigator observed at the fire scene.

The Fire Scene Production

The story of the fire scene video concerns the documentation of how the fire investigation is conducted. Therefore, the camera operator should use the same procedures that are used for documentation with still photography.

The Information Face Sheet

The first shot contains a close-up shot of the information face sheet of the photo log. This is called the introduction shot. It is important to give enough video time so that the information on the card (photographer's name, case number, fire scene location, date and time of the video) can easily be read. When shooting the card at night, be careful not to have a strong light reflect off the card, back into the camera's lens, causing white out. This is also known as wash out. This sequence will be one of the few times when the long and middle shots are not used.

Exterior Sequence

Following the information card sequence, the next series involved the recording of each of the four exterior elevations of the fire scene. The first long shot taken should be an exterior elevation, containing the building number, any name to identify the building, or the registration numbers of a vehicle, aircraft, or watercraft. If none of these items can be found, then find the most permanent identifying feature. The middle shot will give a closer view of the identifying feature and will begin to focus on any evidence of fire damage and/or indicators of how the fire started. The close-up shot brings into clear legible view the identifying features and fire damage. The exterior close-up should also bring into view all exterior openings, windows, and doors, and their relationship to the fire damage. Also, note any signs of forcible entry on the exterior openings. Remember to include other pieces of physical evidence that may be pertinent to the investigation. The same procedure should be applied to the other three exterior elevations.

Interior Sequence

Before entering the building, pick the exterior opening that is going to be used to gain entry into the interior. Video the point of entry in a long shot, followed by a tight middle shot. At the threshold of the exterior opening, video the immediate interior area. Video first the complete interior as it was found by the investigator before debris removal.

After determining the area of least damage, start the interior sequence in this area. Long and middle shots should be enough to document personal contents, furniture, and fire damage when present. If physical evidence is discovered, use the close-up shot to document the item.

Continue to video each room or area using the long and middle shot as the investigation advances toward the area of origin. When physical evidence or fire damage needs to be documented, then do so with the close-up shot. When the area of origin appears to be at the middle of the structure, then start at one end and work towards the other.

In the area of origin, take the time to document with the long, middle, and close-up shots. Pay particular attention to the heat source or physical evidence that has been uncovered during debris removal. Any evidence uncovered should be documented again through the use of the three shots combination, ending with the close-up shot.

The sequence of the interior should include the reconstruction of the fire scene.

Fire Fatality Sequence

In a fire fatality, video first the exterior then the interior as described above. Once the body has been located and documented with notes, measurements, and still photography, then video the location of the body. Video the body from all sides using the long shot to show the relationship of the body to the surrounding area. Use the middle shot to bring the victim in closer and the close-up shot to show the victim's injury either from the fire or sign of a homicide, type of clothing if worn, or any other pieces of physical evidence related to the location of the body. Also, video a sequence revealing the underside of the body and the protected area under the body.

VIDEO IN THE LEGAL SYSTEM

Videotapes used in the legal system are a recent technological innovation that have been in civil and criminal courts since the late 1960s. As a form of moving pictures, videotapes as evidence have taken the place of motion pictures due to availability, cost efficiency, and capability for producing instant results.

Going into the 21st century, the courts have realized that more and more videotape evidence will be entering into the legal system from many different avenues. To establish a coherent system, videotape evidence has been categorized into three sections: contemporaneous videotape, immediately subsequent videotape, and retrospective videotape.

Contemporaneous videotape is the filming of actual facts or events that are recorded live. The videotape will be in controversy due to the camera capturing what actually happened. A good example of contemporaneous videotape involves the recording of an actual crime in progress. Two recent cases captured by television cameras illustrate this category. On April 19, 1991, riots broke out in Los Angeles when police officers were acquitted in the trial of the videotaped beating of Rodney King. During the incipient stages of the riots, a television camera crew, flying in a helicopter, filmed the removal and beating of truck driver Reginald Denny. The second example of a contemporaneous videotape occurred on January 17, 1993, when a Florida television camera crew filmed an interview at a North Lauderdale cemetery. During the interview, a man approached behind a television news reporter and his ex-wife, pointed a 9mm handgun behind the ex-wife's head, fired the gun, and continued to shoot 11 more times as the woman laid on the ground.

Immediately subsequent videotape is the filming of facts or events immediately after or soon after the facts and events of controversy occurred. An example of immediately subsequent videotape would be the filming of a fire scene, after fire suppression activities, to document the fire investigation. Another example is the video confession of an individual after being arrested.

Retrospective videotape involves the filming of events, statements, or a reconstruction of the controversy long after the facts or events occurred and are still relevant to the controversy. An example of retrospective videotape involves the filming of an reenactment of how a particular fire was initiated. Another example would be the videotaping of a fire investigator's deposition.

After determining the category of the videotape, the next decision made by the legal system involves the video's evidentiary classification. Videotape evidence connected in a fire investigation can be classified as nonsubstantive or substantive.

Nonsubstantive evidence is used as demonstrative evidence to illustrate the testimony of a witness. The old school of thought was that any type of photographic evidence was a form of pictorial communication, nonsubstantive, and therefore, did not contain probate value.

Substantive evidence is used to help prove a fact at issue in the trial through two forms of evidence, original real evidence and direct evidence. Original real evidence is tangible and plays an actual part in the event, such as a murder weapon. Direct evidence is used to prove the existence of a fact, such as a contemporaneous videotape. Both types of evidence have probate value. An example of substantive evidence is a video shot as a silent witness.

Lastly, the legal system, as in photography, developed two theories concerning the portrayal of videotape evidence. The first one is the traditional view of all photographic evidence being pictorial communication. The second concept deals with the silent witness theory. As stated earlier, the pictorial communication theory deals with photographic evidence that is used to illustrate the testimony of a witness. The silent witness theory deals with the camera and film capturing an event either observed or not observed by a witness, without the use of a camera operator. For example, a surveillance video camera may record a crime taking place or evidence of a crime, minus the camera operator. Another form under the silent witness theory is the filming of an reenactment of a crime using actors or the actual individuals involved. What is observed on the film does not need an explanation as the event in question, the film "speaks for itself."

FOUNDATION OF ADMISSIBILITY

As more and more forms of pictorial communication were used in the courtroom, the less stringent became the foundation of admissibility. By the time videotape evidence became an instrument of evidence, its foundation of admissibility had decades of photography, motion pictures, and audio recordings to build upon.

At one time, before a photograph or motion picture could be admitted into evidence, the witness had to testify, in part, to the following:

1. The technical skill and education of the photographer.
2. The technical detail of the camera, its position at the scene, the type of lens, type of film and type of lighting.
3. The type of equipment used in developing the film and how the film was developed.
4. The type of equipment used to project the film for the trier of facts.
5. A complete account of the evidence-chain of custody.

When sound motion pictures and other forms of audio recordings became available, some jurisdictions developed a separate foundation called the seven-pronged formula. In 1955 the court case *Steven M. Solomon Jr., Inc. v Edgar* (92 GA. App. 207, 88SE 2d 167) became the first case to establish the seven-pronged formula that stated.

1. That the recording device is capable of taking testimony.
2. The operator of the device is competent.
3. The recording is authentic and correct.
4. That changes, additions, or deletions have not been made.
5. That the recording has been in a complete chain of custody.
6. That the speakers are properly identified.
7. That the testimony was voluntary and made without improper inducement.

Pictures with sound added a new dimension to the admissibility of motion pictures and videotapes as evidence. The audio and visual components are considered separate items of evidence. The rules of admissibility must be met for audio, under the admissibility for sound recordings, and visual, under the admissibility for photographs and motion pictures. It is possible for the court to exclude either the visual or the audio, thereby forcing the foundation of admissibility to rest with the remaining item of evidence. For example, a videotape has been submitted as evidence. The visual, for whatever reason, has been found inadmissible and the audio admissible. The seven-pronged formula may now come into action in order to qualify the audio segment as evidence.

The judicial system began to relax these stringent foundations on the philosophy that the judges and juries throughout the United States were becoming disinterested in the time-consuming technical details of the demonstrative evidence. What was important was the fact that the evidence was a fair and accurate portrayal of what actually happened through the testimony of the witness.

With the advent of video the courts have stated in many cases that videotapes are closely related and can be used in court under the same rules and principles as still photographs, motion pictures, and sound recordings.

One of the first cases involving the admissibility of videotapes occurred in the 1969 case *Paramore v. State of Florida* (299 So 2d 855). In this case, a Mr. Paramore was convicted of murder. At the trial the jury was allowed to see a videotaped confession of Mr. Paramore. On appeal to the Florida Supreme Court, Mr. Paramore requested a retrial based in part on the grounds that the videotaped confession should be excluded due to the state's failure to establish a continuity of possession of the "easily alterable tape." Justice Adkins, in his opinion, stated:

> It appears from the evidence that the video was an accurate reproduction of the entire interview between the officer and the appellant. This being shown, it was not necessary for the state to prove a continuity of possession in order to have the video tape admitted into evidence. This rule governing admissibility into evidence of photographs applies with equal force to the admission of motion pictures and video tapes....We concluded that the videotaped confession was properly received in evidence.

In another court case, *State of North Carolina v. Cornelius Cannon, David Lee Redmond* (374 SE 2D 604), the opinion of the court brought together the different attitudes of what consisted of videotape evidence. First, the court stated that "videotapes are admissible into evidence for both substantive and illustrative purposes." Second, "videotapes should be admissible under the rules and for the purposes of any other photographic evidence." Lastly, a proper foundation of admissibility can be met by:

1. Testimony that the motion picture or videotape fairly and accurately illustrates the events filmed.
2. Proper testimony concerning the checking and operation of video camera and the chain of evidence concerning the videotape.
3. Testimony that the photographs introduced at trial were the same as those (the witness) had inspected immediately after processing.
4. Testimony that the videotape had not been edited and that the picture fairly and accurately recorded the actual appearance of the area photographed.

In many jurisdictions, this court case further relaxed the foundation of admissibility for videotapes by:

1. Making the videotape evidence either nonsubstantive as pictrial communication or substantive as in the silent witness theory.
2. Elimination of the strict technical foundation requirements, such as type of camera, film speed process of development, and the skills and education of the photographer.
3. Elimination of the seven-pronged formula from most of the jurisdictions.

Even though the foundation of admissibility have been relaxed in most jurisdictions, it should be remembered that during any trial, the court may allow the return of the traditional foundation requirements. This may occur when new technology or innovative procedures are used, when specific questions are developed concerning duplication, editing, or when either the visual or the audio portion of the tape is found inadmissible. Lastly, the attorney may use old foundations in specific cases to help illustrate a particular issue in the case. Therefore, it is recommended that the fire investigator review the different foundations as part of his or her preparation for the courtroom testimony.

As for any form of demonstrative evidence, the admissibility of videotape is up to the discretion of the judge. Each decision is based on the facts and circumstances of each individual case. The judge may order to see the videotape first, without the presence of the jury, to determine the evidentiary value of the film. Or, the judge may view the videotape at the same time as the jury. The judge may ask questions concerning the film's authenticity, what the film depicts, how the film was taken, and under what conditions, before deciding the admissibility.

Essentially, the foundations for videotapes to be admitted into evidence are similar to photographs. They must be relevant, involve some issue in the case, and be properly identified and verified.

Under the relaxed requirement, the courts have allowed a foundational witness through testimony to identify the subject matter and then authenticate the videotape. The witness will testify that what is observed on the tape is a fair and accurate reproduction of what actually transpired during the time the camera was filming the event. (Fair and accurate may be replaced with the phrase true and correct likeness or correct and accurate likeness.) The witness will then testify that the film has been in his custody and at no time has the videotape been edited, rearranged, or fabricated. The foundational witness can be the camera operator, someone who was present and involved in producing the videotape, a person who was in the videotape, or a bystander who observed the events as they were recorded.

The foundations of admissibility for videotapes under the silent witness theory are slightly different. To lay down a proper foundation a witness may be asked to testify to:
1. The installation and view of the camera.
2. The camera's type of operation.
3. Maintenance of the camera.
4. The periodic testing of the camera.
5. How the film was removed.
6. The chain of custody.
7. How the camera activates?
8. Testimony of the witness that the camera recorded accurately the events that occurred.

There are four ways that a videotape can be verified under the silent witness theory. First, testimony from a photographic expert that the videotape has not been altered in any way. Second, the witness can testify to the chain of custody of the videotape. Third, testimony of a witness that the videotape and camera operated and had been properly maintained. Last, testimony to the fact that the videotape at trial is the same observation the witness had observed during the recording.

The fire investigator is not limited to using the video on just the fire scene. The video can be used to document statements, confessions, recorded experiments and crime re-enactments. *A WORD OF CAUTION*–In addition to the foundations of admissibility for a video recording, each specific subject will have a set of rules to follow.

The leading court case, *Hendricks v. Swenson* (CA 8 MO, F2d 503), involved the use of a videotape recorded during a statement/confession. The defendant felt that his Fifth Amendment constitutional right against self-incrimination was violated when his voluntary confession was videotaped and shown to the jury. According to court testimony, the defendant was told he was being videotaped, recording both his picture and voice and that it would be used against him in court. The court stated:

> .....we suggest that a video tape is protection for the accused. If he is hesitant, uncertain or faltering, such facts will appear. If he has been worn out by interrogation, physically abused, or in other respects is acting involuntarily, the tape will corroborate him in ways a typewritten statement would not. Instead of denying a defendant his rights, we believe it is a modern technique to protect a defendant's rights.

Judge Van Pelt, of the 8th Circuit Court of Appeals, also approved the use of videotaped confession. Judge Van Pelt stated:

> If a proper foundation is laid for the admission of a video tape by showing that it truly and correctly depicted the events and persons shown, and that it accurately reproduced the defendant's confession, we feel that it is an advancement in the field of criminal procedures and a protection of defendant's rights. We suggest that to the extent possible, all statements of defendants should be so preserved...For jurors to see as well as hear the events surrounding an alleged confession or incriminating statement is a forward step in the search of the truth.

Further cases have ruled that if a defendant was given his full Miranda Warnings and had waived those rights, a videotape of the defendants statement could be used even if the defendant was not advised a recording was made, both audio and visual of the event.

Another important detail to be considered during the videotaping involves what is in the background. Many times the background unintentionally will contain items (such as religious pictures or models of an electric chair) observed during the confession of a murderer. These and other items may cause the videos to become inadmissible by being prejudicial towards the trier of fact's decision. It is best then to have a neutral background with only the individual being interviewed and the interviewer in the picture.

Under no circumstances does the interviewer promise the subject that the videotape will not be used against him or her in any type of criminal proceeding. Like all other types of confessions, any promises made can cause said confession to be inadmissible.

Videotaping an experiment follows the same foundation for filming an experiment with motion pictures. The video is used when it may be dangerous to perform the experiment in the courtroom. The experiment may be too large or difficult to move into the court. An expert should be the one conducting the experiment. In this way the expert can testify and give his or her opinion as to what the video is depicting. For the video to be admissible, it must be relevant, be a fair and accurate record, be helpful to the jury, and it cannot be prejudicial.

Reenactment of a crime can be acted by the defendant or by individuals other than the defendant. When using the defendant in the reenactment, the video becomes a pictorial confession with the foundation requirements being consistent to those for a video confession. After documenting the identification information, the next important recording is the audio and visual giving and response of the Miranda Warning.

Reproducing the crime is a delicate simulation in the eyes of the court. The reenactment should be conducted under similar circumstances of the original event. Differences between the original and the reenactment may be confusing to the jury and variation may be deemed inadmissible by the discretion of the judge. Some examples:

- Original event occurred during the night and the reenactment was filmed during the day.
- The original weather is different in the reenactment.
- The furniture is different in the re-enactment.
- The outdoor reenactment is different from the original.

Many courts do not encourage the use of actors when filming the recreation of the crime scene. A Texas court case *Lopez v. State* (651 S.W. 2d 413) gives a general view of how the court feels:

> While videotape re-creations of criminal activities may be acceptable in some jurisdictions, the concept of re-creating human events with the use of actors is

a course of conduct that is fraught with danger. The general appearance of an actor his facial expression, or slightest gesture whether intended or not, may sway jurors who have listened to lengthy testimony. The danger of jurors branded with television images of actors, not testimony, is too great to ascertain. No court instruction could remove highly prejudicial evidence of a re-enacted rape or murder if we establish this precedent.

Before embarking on the filming of a reenactment, it is best to consult your state or federal prosecuting attorney to see if such a video would be admissible.

When the videotape is completed and what is captured on the film meets with satisfaction of the investigator, remove the tab on the back of the plastic case. This will prevent the tape from accidentally being recorded over, destroying the original pictures. Then fix a label to the videotape with the following information:
- Description of video—fire scene, confession, etc.
- Address at which video was taken.
- Date/time video was taken.
- Case number.
- Name of camera operator.
- Number of feet used/running time in minutes.

There are several objections that an opposing attorney could use to have the videotape inadmissible. First, a question may be brought up that the videotape does not prove or disprove an issue in the case, and therefore is not relevant. Second, the videotape will confuse the trier of fact. Third, the videotape may be more prejudicial than its weight as evidence. Lastly, how the videotape was produced, how it was stored, who the camera operator was, and other pertinent information may be an issue.

FINAL VIDEO TIPS

- When shooting video of a fire scene or a collapsed building the camera operator's safety is important. When looking through a viewfinder the vision is restricted. Learn how to shoot with both eyes open, one eye in the view finder and the other looking ahead. Practice will overcome this unnatural position.
- Wear protective clothing, hard hat, gloves.

- Plan route of movement ahead of time.
- Be aware of weak floors, electric wires, protruding nails, sharp objects, fallen debris, and overturned furniture.
- Look at the scene before shooting pictures. In your mind, determine what it is you are trying to capture on the film.
- Always carry information cards like in still photographs. Shoot the information on the card before anything else (address, person taking the video, date/time, and case number).
- Whenever possible, shoot at eye level, so that the scene will look as normal as possible. This will give the viewer the sense of actually being there.
- Most video cameras have a time and date indicator that will record said information on the tape. Use this at the beginning of the tape. If there is a break in time, over several minutes, document, in writing, explanation of the time delay. (Defense attorneys may question the missing time interval.)
- Get into the habit of reading your indicator lights within the viewfinder. These lights will tell the camera operator when the camera is recording or not, how much tape is left, how much battery is left for operating and the position of the zoom lens.
- Use a tripod whenever possible. This will stop jerky movements and vibrations that will distract and distort the view.
- Do not walk and pan the camera at the same time. Along the same lines, do not walk and shoot at the same time unless you have the technique and knowledge to do so.
- Once you have established the basic procedures of videotaping a fire scene and are comfortable, stick to the routine on each fire scene that you do. This can eliminate any bias opinions brought up by defense attorneys.

Before shooting a fire scene, practice and practice some more. Your first few times shooting should not be a fire scene. Instead, get familiar with your camera, study the operating manual, review what they can and cannot do on any other subject matter. Critique your tapes and ask yourself different questions for self-improvement: What should I have done for a better picture? How can I improve the quality of the video?

Through the use of video, a fire scene comes to life through motion and sound and records the event in a measure of time. To become part of the courtroom presentation, the fire video must pass several tests in order for it to be introduced into evidence:

1. It must function properly both in the visual and audio version. The visual should be clear and the audio audible.
2. That the camera operator knows how to operate the camera. Not as an expert but someone experienced.
3. That the visual and audio are authentic and correct.
4. That there were no additions, deletions, or changes made after the initial recording.
5. That the videotape was properly preserved.
6. If the video is a confession or statement, it should be made voluntarily and without promises of any kind.

Used in conjunction with field notes, sketches, and photographs, video is another form of demonstrative evidence that aids the fire investigator in documenting the fire scene.

Chapter 11

FIRE SCENE MODELS

As preparation begins to bring the case to trial, a decision may be reached that sketches and photographs are not enough to help illustrate the testimony of a witness. The uncertainty focuses on whether or not the two-dimensional plan of the sketches and photographs can clearly communicate to the trier of fact the verbal descriptions made by the witness. A solution to this dilemma is through the construction of a fire scene model.

As one of the simplest forms of demonstrative evidence to comprehend, the fire scene model is a three-dimensional presentation aid that reduces the original size of an object, a location, a form of movement either manual or mechanized, or an idea. The fire scene model enables the trier of fact to observe and understand the relationships, movements, or concepts of important issues in a case. The model gives a complete view of the miniature, from every angle that can be seen by everyone at the same time. The small scale representation can be a working model, revealing working parts, or a stationary model showing location or appearance.

The legal model must be a true and accurate representation of the original. Models attract and will hold the attention of the trier of fact. It seems that models have a magical fascination that the viewer cannot resist gazing upon. The model is viewed as a simple form of communication that is an aid in a better visualization compared to technical drawings or photographs. Even with the use of the computer, a computer-imaged model cannot replace the actual material experience, the physical shape, and the relationships between each of the three dimensions. The modern juror has learned to use this visualization through the use of television. Jurors think better by seeing. Once the triers of fact sees the model presentation, they are attracted to and their attention span is held as the convincing points of evidence are made. Standing by themselves, crucial points made in verbal testimony may be misunderstood or go in one ear and out the other as the attention span wanders. As the trial continues into days, the trier of fact will unlikely overlook or forget a point that had been made with the model presentation.

If at all possible have the model present during the opening statements of the trial. Especially in a jury trial the model will be observed when a point is made about the original in the statement. After opening statements the model will remain in the courtroom throughout the trial as a visual aid for the jury. At closing arguments the model will be present as the important points are reiterated, leaving a lasting presence in the juror's mind. Lastly, if the model was admitted into evidence, then the model may be brought into the jury room while the jury is making their decision.

There are several differences when comparing an architectural model and a legal model (i.e., a fire scene model). The main purpose of the architectural model is to use the model as a study of spatial (space) relationships of structures and land topography. The model provides a means to observe, test, and evaluate designs before the actual building construction begins. Once the design is approved the model is used as a presentation tool to sell an idea. The fire scene model is used only as a presentation tool. Finally, by reducing the size of the original object or place, the fire scene model can take the place of the original that otherwise would be too large or impractical to transport into the courtroom. Therefore, the legal model will be the type used by the fire investigator.

ADMISSIBILITY OF THE FIRE MODEL IN COURT

As with any other form of demonstrative evidence, the determination of the admissibility of the model begins with the judge. Through the testimony of a witness, a model must be identified and verified as being a fair and correct representation of the origin, that the witness is qualified to speak about the model and that the model is used to illustrate the testimony of the witness. Once this foundation has been accomplished, the judge will then decide the relevancy of the model, if it contains prejudices, if it is going to mislead the jury, and if the model is a part of cumulative evidence.

By reviewing some of the court cases involving the use of a model, one will begin to get a better understanding of what the courts generally expect when a model is presented as demonstrative evidence.

Throughout the judicial history of the United States, models have been admitted as evidence, as a form of demonstration or illustration, through the discretion of the judge. In one court case, *Church v. Headrick & Brown*, (101 Cal App. 396, 225, P2d, 558) the court stated that the admission of a model was largely with in the discretion of the trial court.

One of the best court cases to address the trial court's discretion is in *W. R. Smith v. Ohio Oil Company* (10 Ill. App. 2d 67, 134 N.E. 2d 526). Justice Scheineman of the Appellate Court of Illinois, Fourth District stated:

> ...the use of such [demonstrative] evidence is usually left to the discretion of the trial court, and expressions of disapproval are generally based on irrelevance, or that the model, picture, etc. was misleading or not explanatory...

> ...if it appears that the exhibit was used for dramatic effect, or emotional appeal, rather than factual explanation useful to the reasoning of the jury, this should be regarded as reversible error, not because of abuse of discretion, but because actual use proved to be an abuse of the ruling.

The courts also realized that by using models, a clearer image of the testimony may be seen by the trier of fact. In the court case *Reinke v. Sanitary District* (260 Ill. 380, 703 N.E. 236), the court felt that the use of models would give the trier of fact better information on places and objects that the jury would otherwise not have seen. Many models are not evidence by themselves and are used so that the trier of fact will have a better understanding of the witness's testimony.

In another court case, *William F. Sherman v. City of Springfield, Illinois* (250 N.E. 2d 537), Justice Thomas J. Moran of the Appellate Court of Illinois, Fourth District, stated:

> ...our courts have long followed a procedure of allowing working models of machinery and mock-ups into evidence when their admission tends to further clarify a subject or question in issue.

With more models and other forms of demonstrative evidence finding their way into the judicial system, it became common practice for the model to be identified as a correct likeness of the original and introduced as part of a witness' testimony. As early as 1856, in the case *State of New Jersey v. Fox* (25 N.J.L. 566,602), the courts felt that a model is a copy of the original and that the witness must state that the model is correct. Judge C.J. Green stated:

> A model is a copy or imitation of the thing intended to be represented and when the witness states that he exhibits a model, it is to be inferred, in the absence of all proof to the contrary, that the model is (sworn to be) correct.

In another court case, *Hassam v. Safford Lumber Company* (82 Vt. 444, 74 A 197), the court went a step further and stated that not only is the model to be an accurate representation of the original, but the witness must verify that fact. Simply put, the witness must have personal knowledge of the original before testifying to the facts. The court stated in their decision:

> Models, maps, plans and photographs belong, in the law of evidence, to the same class, and are admissible only when properly verified; that is to say, preliminary evidence is required to show that they are sufficiently accurate to be helpful to the jury. They are not to be received anonymously, but someone, not necessarily the maker, must stand forth as their testimonial sponsor. But, this preliminary question of the sufficiency of the verification, through a question of fact, is for the determination of the court, and is not ordinarily reviewable.

In 1984, in the case *Gallick v. Novotney* (124 Ill. App. 3d 756, 464 N 2d 846), the Appellate Court of Illinois determined that a model was properly excluded on the foundation that the witness did not give testimony to support the model's accuracy with respect to the critical facts shown.

Another issue encountered in the courts focused on the size and accuracy of the model in relation to the original. The courts all along have felt that a model does not have to be identical to the original in every detail. However, the reproduction should be realistic and accurate enough to compare with the original. Before being admitted, any of the discrepancies between the model and the original must be clearly explained and cannot mislead or cause prejudice with the jury.

A review of some of the court cases will help in understanding how the courts have reacted towards the many discrepancies concerning the size and precision of models.

In the case *City of Tucson v. LaForge* (8 Ariz. App 413, 446 P 2d 692), the court was confronted with a model that was not an exact replica of the original. In its decision on the model's admissibility the court stated:

> Where such use would not tend to mislead the jury, use of a model is permissible even though it is not a facsimile in every detail.

In another court case, *People v. Speck* (41 Ill. 2d 177, 242 N.E. 2d 208), a scaled model of a townhouse was used in a murder trial. The murder scene consisted of eight female victims, with the model used during the testimony of a surviving eyewitness. The model helped the jury understand how the crime was committed. Part of the model contained wooden blocks of seated figures to indicate the location of each victim when they were alive. Another series of wooden rectangle blocks were used to show where the murder victims were found. During the appeal, the defense argued that each wooden block of the seated figures looked like a woman kneeling in prayer and the second series of blocks resembled coffins. Each of these blocks was used to inflame and prejudice the jury. State Supreme Court Justice J. Klingbiel stated:

> It is apparent that the sequence of events and the various locations within the town house would be difficult for the jury to follow without the use of some physical exhibits...In our opinion the exhibits were a useful adjunct to the testimony and were not of such a nature as to be prejudicial.

One of the biggest discrepancies involves whether the model is to scale or not to scale, and if not, is it admitted into evidence? As with any other form of demonstrative evidence, this decision by the trial judge is determined on a case by case basis.

In the court case *Arkansas State Highway Commission v. Rhodes* (240 Ark. 565, 401 S.W. 2d 558), a model, partially not to scale, was admitted into evidence. Part of the appeal concerned the unscaled portion of the model misleading the jury. The model represented a site involved in an eminent domain hearing. The vertical items such as trees, building, fences, and telephone poles were not of the same scale as the horizontal items. During the witness' testimony, he stated that he prepared the model and explained the different scales. Along with the model, several aerial photographs were used. Justice J. McFaddin, on appeal, stated:

The witness who prepared the model, carefully explained the scaled matters. There was nothing misleading. Besides, there were a number of enlarged aerial photographs introduced, and these, along with the model, were designed and apparently did give the jury a thorough understanding of the situation.

In a federal court case, *Burriss v. Texaco, Inc.* (CA 4 SC, 361 F 2d 169), the court found that a reenactment involving a model was inadmissible because it was not to scale. The model, depicting a railroad yard and a feed mill, was prepared by a defense witness. The witness was permitted to give his expert opinion on the origin and spread of the fire. However, when the witness wanted to reenact the fire on the model, he was not allowed. Circuit Judge Sobeloff of the United States 4th Circuit Court of Appeals stated:

> Rejection of the model was particularly justified here as the tubing used to represent the drainage pipes was admittedly not to scale, rendering the conditions of the proposed experiment substantially different from those existing in the fire.

During the early 1960s the court case *State of Washington v. Henry LeRoy Gray* (395 P 2d 490) brought together into one, the different court opinions. Judge Half of the Supreme Court of the State of Washington stated:

> Courts should approach the admission of models, samples and things offered exclusively for illustrative purposes with wariness and circumspection, to the end that fact be not confused with fancy and artistic interpretations pushed aside and take over the role of truth unadorned. A ruling designed to allow the admission of a model house in evidence should guardedly preclude an air castle, even though the two may have cogent similarities. Thus, models, samples and objects offered in evidence for purely illustrative purposes must not only be relevant and material in character to the ultimate fact sought to be demonstrated by their use, but, additionally, must be supported by proof showing such evidence to be substantially like the real thing and substantially similar in operation and function to the object or contrivance in issue. If the proffered evidence does not meet this test it should be rejected.

In review, to have a model as an exhibit of evidence, it must pass the test of admissibility through the testimony of the witness. The witness will lay down the foundation by:

- Stating that the model is a fair and accurate representation of the subject matter or the area in question.

- Stating that he or she is qualified to testify as to the accuracy of the model.
- Stating that the prepared model is identical with the original except for size.
- Stating that the model was prepared according to scale or not to scale.
- Stating, that through the aid of the model, it is used to assist and help illustrate his or her testimony so that the trier of fact may have a better understanding of the testimony.
- The judge will decide if the model will mislead, confuse, or cause undue influence to the trier of fact.

TYPES OF FIRE SCENE MODELS

The extent of fire scene models can vary from the instrument that caused the fire to the large plot plan covering several buildings. The majority of the models for courtroom presentation will be the following architectural types:

Structures

Architectural building models will only show the building standing by itself. The purpose of this type of model is to display the building's different components, such as roof lines, size of the building, color of the building, location of exterior doors and windows, and location of important objects within the curtilage. Plant foliage may be omitted unless the plants are important issues in the case. For example, Figure 11-1 are photographs of an actual structure. Figure 11-2 shows how a model can represent an actual structure. Notice that the distractions such as utility lines and landscape vegetation are eliminated, giving the model a clear overall view.

Figure 11-1. A single story structure (photograph).

Figure 11-2. A structural model representing the building in Figure 11-1 (photograph).

The Floor Plan Model

When the complete structure is at issue, then the floor plan model is constructed to reveal how the rooms relate to one another and the location of any evidence found. The exterior walls will reveal the structure's elevation on each exposure. The model will not have a ceiling and may or may not have a roof. Other objects of interest may include furniture, appliances, location of interior and exterior openings, location of hallways, foyers and stairs.

To illustrate this point an actual floor plan fire scene model is shown in Figures 11-3a-d. The fire scene was a single story, wood frame, one family dwelling. The fire involved two fatalities along with three severe fire injuries. A suspect was arrested and charged with two first degree murders, three attempted murders, first degree arson, and burglary. In preparation for the trial, a meeting was held with all the fire investigators and the Assistant State Attorney. All parties agreed that a model would help illustrate the testimony of witnesses.

Figure 11-3a. The floor plan model (photograph).
Front elevation.

Figure 11-3b. Rear elevation (photograph).

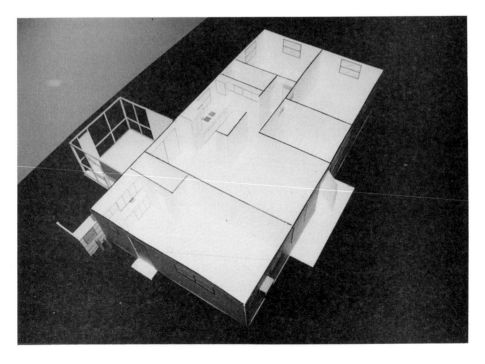

Figure 11-3c. Model with roof removed (photograph).

Figure 11-3d. Detail close-up of the model's kitchen (photograph).

The three-dimensional model's exterior and interior walls were constructed of foam board with a sheet of plywood as a base. The model was built with three functions in mind. First, in conjunction with photographs and sketches, the model would give each jury a vivid impression of actually being at the fire scene. Second, due to the victims being found in different areas of the interior, the investigators felt the jury could easily get confused in how each victim was located. To overcome this problem, the model was designed to have the roof removed (Figure 11-3c) to expose the interior floor plan. Hearing and viewing the three-dimensional model during the fire fighters testimony, the jury could orient themselves to the fire scene, travel the same routes as the first-in fire suppression personnel who entered the burning structure and learn how each victim was located. Third, the rear of the model was constructed to reveal the area of origin in detail. Through this documentation, each fire investigator could illustrate the investigation process and explain how the fire classification was determined. (A side note: Before the conclusion of the trial, the suspect pled guilty and is currently serving his sentence.)

Diorama–A One Room Model

To capture the interior of a room within a model presents a problem for the viewer. Since the model is a miniature representation of the original, there must be another way to physically place the viewer inside the room. Whatever technique used, care must be taken not to disrupt or destroy important features. To resolve this problem, the room is constructed into a diorama. The diorama cuts away one of the walls and ceiling so one can look directly into the room and its furnishings. Transparencies can be used to take the place of walls and ceilings so that they may be used with a grease pen for making important points. The viewer now has a complete view of vantage points and important objects that would otherwise take many photographs to present the same area. The diorama also brings out the three-dimensional relationship between the structural members and the furnishings that cannot be found in the two-dimensional drawings of the floor plan. One of the benefits of the diorama is that heat and smoke patterns may be integrated into the model to show fire travel.

Figure 11-4. A residential living room (photograph).

Figure 11-5. A diorama model of the living room in Figure 11-4 (photograph).

To illustrate this point, Figure 11-4 is a photograph displaying a residential living room. Figure 11-5 is a photograph of a constructed diorama model of the living room in Figure 11-4. Note how stylization simplifies the photograph by eliminating unimportant items.

Construction of the Diorama Model

The architectural interior model is used to study design, room function, and the best possible location of furniture. This type of model is constructed from blue prints and designed with a purpose. The fire scene interior model is a reconstruction of an already existing room with furniture. Like any other type of legal model, the fire scene model's purpose is to bring the original, in a smaller version, into the courtroom.

The building of a fire damaged room into a model introduces a number of unique problems. To begin a successful interior model, the investigator must take detailed notes, sketches, and photographs of the area of origin. Remember, when observing a fire scene, many of the physical characteristics of objects have been altered by the fire and fire suppression activities. An investigator who is a novice, in a hurry to get home, wants out of the harsh climate, or for whatever reason, may overlook the modified object that will be crucial to the model. Make it a practice on every fire scene to document:

- The location of furniture and personal effects and the extent of damage inflicted.
- The shape and dimension of the area.
- Type of construction.
- Material composition of interior walls, floor, and ceiling.
- Use colors similar to the original. This may sway the witness to state that the model is an accurate representation of what the area looked like prior to the fire.
- The construction and composition of the point of origin.

Further detailed information can be developed through interviews of occupants and/or owners. Witnesses can describe what was inside the room. If the room is badly damaged, always try to have an occupant draw a sketch of the structure and the location of interior doors and furniture. This is usually done before removal of any fire debris. The best witness to have recollection to draw the sketch is the one who stays home the most and/or does the house cleaning.

The investigator who gathers as much information as possible of the fire scene, no matter how minute, will bring out in greater detail a realistic model.

To put together a diorama first decide which interior wall is the least important. This will be the wall that is removed to reveal the interior. Using the desired scale, layout the floor plan on either foam board or illustration board. The next step is to construct the floor resembling the actual building material. This can be obtained in several ways:

1. Drawings–many types of flooring can be portrayed through pen and ink drawings, colored with felt-tip markers. With a little practice flooring such as field stone masonry brick, tile, and wood can be made. This technique can also be used on interior and exterior walls.

2. Photographs—taking pictures of the actual material then applying the photographs to the floor or wall. The big disadvantage is finding the appropriate scaled photograph to go with the model. Also, the photograph may be glossy compared to the dull finish of the actual material.
3. Use of a material that resembles the original material. Gray cardboard can be used as concrete. Cloth can simulate carpets. Small area rugs made from cardboard and colored with felt-tip markers.
4. Model making kits found in many hobby stores, building materials can be found that are used to construct doll houses and model railroad scenery.

The list is by no means complete. Innovative ideas will produce new model materials for both the old and newly introduced building construction material.

Furniture can be made as a stylization concept. The most realistic stylization is the furniture constructed from pieces of foam board, followed by illustration board, to blocks of wood. Furniture dimensions can be found in architectural textbooks, interior design books, and furniture catalogs. The dimensions of the particular pieces of furniture being reconstructed is made in the same scale as the model. Once cut, the pieces can be given color and texture. The final touch is gluing the pieces together.

Complete the constructions of all the furniture needed for the diorama. With the floor completed, all the furniture is then placed in its prospective position. Before gluing, make certain all pieces are properly placed according to sketches, notes, and photographs,.

The interior walls and ceiling are cut to dimension and detailed with fixtures (if applicable) and wall texture. The walls are placed together, adjusted until correctly in place, then glued.

If the model maker is proficient in using a circular and a band saw, furniture can be constructed with greater detail. To save time, furniture can be bought from stores that sell miniature furniture for doll houses.

Sectional Model

This type of model is based on minute detail and measurements. There are many variations to this type of model. First, the model can be a part of the whole, as was the case in the electrical outlet in the interior wall of the *BEVERLY HILLS FIRE LITIGATION*. Second, in

a few of the cases, the model can be an exact replica of the original
such as an ignition device or a pipe bomb. Third, the model may have

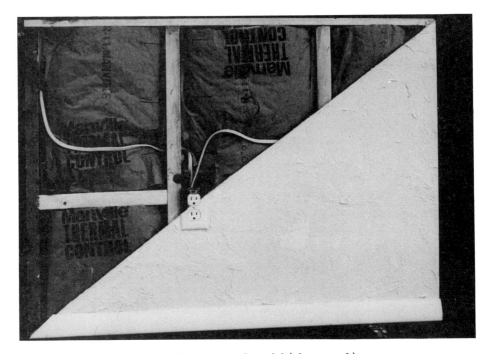

Figure 11-6. A sectional model (photograph).

a section taken out to expose the working parts, or the point of origin, that are hidden or places difficult to observe. This is called a cutaway. Figure 11-6 is a sectional cutaway of a wood frame wall involving an electrical outlet.

Neighborhood Model

The neighborhood model or site plan should show the important
objects within the fire scene and how they fit into the surrounding
area. The relationship of buildings to one another and the location of
evidence can be shown. Figure 11-7 is an example of a neighborhood
model that was constructed for a criminal federal court case involving
a militant religious group. The fire scene involved a residentical city
block where over 75 Molotov cocktails were thrown. Six houses were

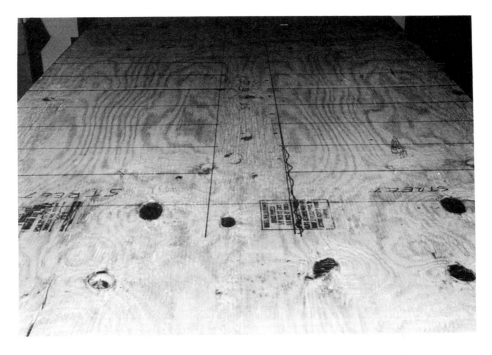

Figure 11-7a. Construction of the model's base (photograph).

Figure 11-7b. Completion of the model's base (photograph).

Figure 11-7c. Aerial view (photograph).

Figure 11-7d. Aerial view (photograph).

damaged by the fire bombs. Three fire injuries were sustained including a ten-month-old baby, whose crib ignited in flames when a bottle exploded on the wall above the crib.

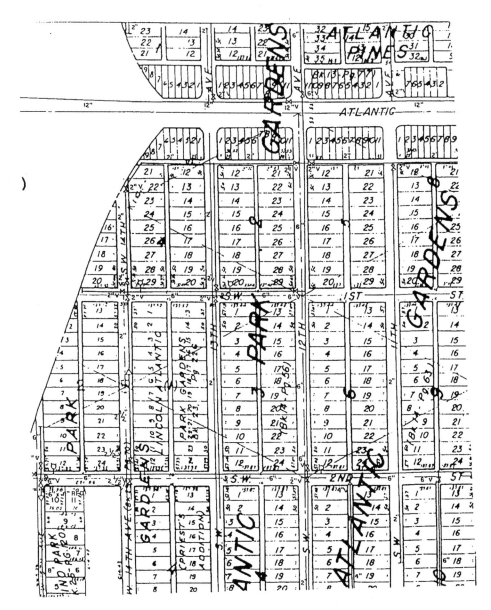

Figure 11-8. A city engineering plan.

Figure 11-7a shows the beginning stages of the model construction. Three-quarter-inch-thick plywood, because of its strength and rigidity, was used for the base foundation. On top of the plywood a property lines was laid out according to a city engineering plan in Figure 11-8.

Figure 11-7b reveals the completion of the ground cover. Grass was made using grass paper from a hobby shop (train modeling department), tar paper for the asphalt road, gray cardboard for the concrete driveways and sidewalks, and aluminum window screen with small finish nails for the chain link fence. Figures 11-7c and d gives the viewer an aerial perspective of the neighborhood.

Topography Model

The purpose of a topography model is to show a site of property surrounding a building or neighborhood. In this type of model, topography, the elevation and terrain features, may become important details in the model. Other objects in this model may include driveways, foot paths, fence lines, natural ground texture (such as sand, grass, rock formations), water surfaces, relationship of buildings to prominent landmarks, telephone and electric poles, and vegetation.

Construction of a Topography Model

The simplest form of topographic model would be a flat area of land. Like the survey markers on the construction site, the model's design is transferred onto the base plate from the different sources of information gathered. First, layout all the roads, driveways, sidewalks, and paths. If water areas are involved in the model, they would be the next area added. To eliminate the confusion of many different lines, the water areas are shaded or outlined in blue. A darker shade or the outline of green can be used to distinguish the greenery areas. The position of the structure(s) are then placed in their proper locations on the base. After all markings are completed the water areas are assembled first. Next comes the green areas, then the roads, sidewalks, and so on. Structures are the last major items fastened to the base. The remainder of the work involves the placement of detailed objects to show the proportionate ratio of the model.

If the original site or property contains variations in elevation and shape of the land, it may be necessary to incorporate these features into the site model. To create a topographic model, a site survey plan from the mortgage holder or survey company can be obtained.

At a quick glance, the topography map looks confusing. With elementary information, a model builder can take the jigsaw puzzle of lines and transform a topography plan into a topographic model.

An example of a site survey plan is seen in Figure 11-9. The property lines are distinguished as thick solid lines, medium length with two dashes separating the solid lines. Each leg of the property line is described by direction or bearing, then length. Direction is determined by the compass direction of true north and south and is at zero degrees. The property line is then determined by the number of degrees east or west of the true north/south line. This angular bearing should be less than 90 degrees at any given time. For example, if the direction on the property line states N 86 degrees E, it indicates that the line is 86 degrees from true north towards the east. The length can be denoted in feet and inches or in the civil engineer's scale of feet and decimal parts of a foot. When a property line intersects another property line, a property corner is established and is shown as a small circle with the height of elevation given. From the corner, the new direction and length of the property line is given. Each part of the perimeter is given until the property line returns to the beginning point. With the completion of the property line, the exact shape of the lot and its dimensions are given.

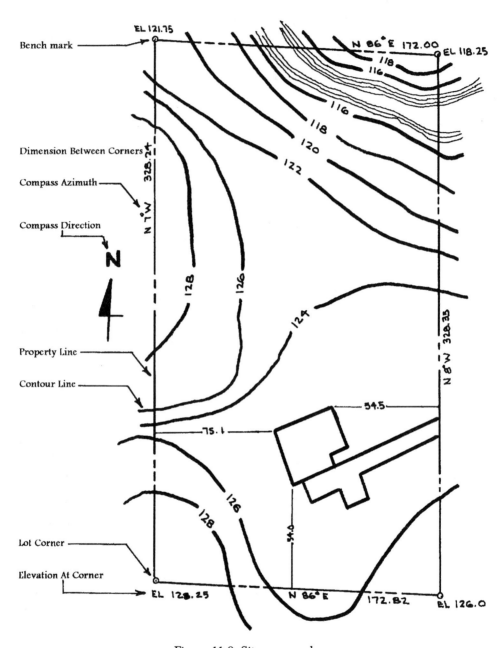

Figure 11-9. Site survey plan.

Elevation is determined by the datum point. The United States Geological Surveys uses sea level as the universal datum point. Some jurisdictions established their own local datum point. In either case, the datum is zero elevation. From the datum point, elevation is shown through the use of bench marks. Elevation is also shown through the use of contour lines that provide a three-dimensional picture of how the land appears in a two-dimensional drawing. Contour lines have two functions. First, contour lines are points of similar elevation that are connected together by the same line. This line represents the height of the land at that particular point. Second, the shape of the contour line describes what the terrain looks like. The spaces between the contour lines are called contour intervals and represent the vertical distance between contour lines. These intervals can range from 1' to 10'. A scale on the plan should tell the reader the interval. If the contour lines are close together, the elevation is changing rapidly and the slope steepens. If the contour lines are spaced far apart or there are no contour lines present, the land is gradually sloping or flat. Each contour line in the plan means a change in the elevation.

Survey drawings will also show the following information that can be helpful to the model builder:
- Complete legal description
- Direction of North
- Scale
- Name and width of the street
- Setback lines
- Easements
- Name of the registered land surveyor

The topography is built up from the base plate either in steps to give an abstract formation of the natural landscape (Figure 11-10), or built as a complex free form. The material should be light and easy to work. Materials such as corrugated cardboard and foam board make excellent stylization for the elevation steps. Clay and wire mesh covered with paper mache give a more natural appearance in the free form. However, the free form model is time consuming and should only be attempted by an experienced model builder.

To create the step topographic model, trace the contour lines, water surfaces (if applicable), and location of structures onto the base plate. If a photocopy is obtained of the survey site plan, cut it with a pair of scissors along each of the contour lines. This will make a pattern of each contour interval. Next, determine how many contour intervals there are and out of the model's material, cut that number into the proper dimensions of the property lines. The lowest elevation will be the ground plane of the base. The contour at the next level up from the ground plane will have its contour lines traced; then cut along the line with the remainder of the board fitting within the property lines. Proceed with the remaining contour intervals in the same manner. Once all the contours have been cut, starting with the ground plane elevation, place all the elevations in their proper position without adhesives. Make sure a proper fit is obtained and all contours are properly cut. If satisfied with the layout, remove all contours and begin the permanent bonding with white glue in the same procedure as before, starting from the bottom up. To prevent the boards from warping, the pieces of material can be held together with pins or held down with weights. Let the layers sit overnight to completely dry the adhesive. The next step is to seal the layers with a shellac so paints can be applied without warping the cardboard. The shellac dries quickly. Next, apply a latex paint to the base, the color being the predominate color of the ground cover.

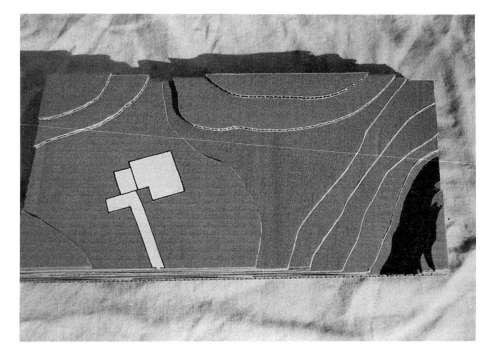

Figure 11-10. A topography model (photograph).

Only use topography models when it is necessary. The construction of such a model can be very time consuming in man hours. When elevation becomes a factor, problems exist when traffic, green, and water surfaces come into play. This type of model should be left to the experienced model builder.

Street and/or Intersection

Making a model of a street and/or intersection involves placing the two-dimensional map onto a rigid board and adding three-dimensional objects (such as buildings, vegetation, vehicles, people, etc.). Most urban streets and intersections are relatively flat. Beware of elevation that could make this a difficult model to build.

Vehicles, Aircraft, Vessels, or Any Other Forms of Transportation

Many fires have gone to court involving these forms of transportation. Hobby stores and mail order catalogues have the various scale models of vehicles, aircraft and vessels. Automobile manufacturing companies may have models of the different engines used in their vehicles.

CONSTRUCTION OF THE MODEL

Before the first line is drawn or the first piece of material cut, a design layout is developed to give an idea of what the model represents. The plan begins with a meeting between the attorney and the fire investigator. Together, ideas are exchanged back and forth communicating a mental image. Keeping in mind that the model will help illustrate the testimony of a witness, decisions are made as to what type of model will be constructed. Is the model going to illustrate a building or a device? If a building is involved, many important considerations need to be analyzed. Is the interior of the structure going to be visible? If exposed, is the interior displayed without a roof or after the roof is removed? With a structure of multiple floors, is each floor removable to expose the lower floor's plan or just the floor in question? If the structure is not important, does the model depict important features of a room or a site plan to show adjacent buildings or prominent attractions of the surrounding landscape? When a model of a device is constructed, it usually will represent the instrumentality of the fire cause taken from reconstructed pieces at the fire scene. Will the model be an exact duplicate as a pipe bomb or a smaller replica of an electrical outlet inside a wall? Another type of model may depict a three-dimensional likeness of the interior or exterior of a structure, showing the relationship of items and/or fire damage found during the investigation. Depending on the purpose of the model, these are just a few of the questions that might be asked.

When the type of model has been established, the model builder should begin to gather as much information on the original as possible. Information will be developed from the fire investigator's field notes, sketches, photographs, and at times, the video. This group will help establish what the fire scene looked like when the investigator conducted his/her investigation. Another group of information can be collected from technical drawings of either architectural or engineering plans and notes from the city or county building departments, and plans and notes from inspections conducted by the fire department.

The remains of homemade devices should be reconstructed, measured, photographed, and notes taken of important features. Look for brand names, model numbers, type of material, etc. used to make the device. This information will ensure the model to being as close as possible to the original.

Once the subject matter of the model has been established, the model is then examined through a series of questions regarding size and the amount of detail needed to illustrate the important points. The model should be large enough so that the trier of fact can easily observe it. If possible, before the actual start of construction, visit the courtroom where the testimony will take place. Make sure the determined size of the model will fit through the courtroom door opening. Inside the room, note the location of the jury box. Directly in front of the jury box is usually where the model will be presented. This will give the fire investigator a good idea on the interaction distance between the jury and the witness. If the model is too large for the courtroom door or too heavy for one section, then divide the model into units. When in sections, the model maker should be present prior to the beginning of the trial to assemble the model correctly.

Another feature of size that needs to be resolved is whether or not to build the model to scale. The scaled model is a fixed ratio of reproduction and should be applied when dimensions, line of sight, and clearances become important issues in a case. The scale will depend on the detail and the appropriate size for optimum viewing. For large structures, the average scale for reproduction is 1/8" = 1'0". This scale will cover a large area in a small space. It is best utilized in construction of a large site plan to display relationship of buildings and/or prominent land features. The loss of small detail is the main disadvantage of using 1/8" scale. For the average size structure the scale of 1/4" = 1'0" will give the best results both in detail and appearance. For smaller structures the model will probably need to be enlarged, thus using the 1/2" = 1'0" scale. For a scale between those mentioned above, an architectural-scaled ruler is an excellent tool to use.

Nonscale models can be used when dimensions are not an issue in the case. Without a scale, the model must be proportionate, thereby making it a fair and accurate representation of the original. Since detail to dimension is not essential, the nonscale model will reduce the man hours in construction, making the model easier to build and will considerably lower the cost of the project.

The construction of the model and its items should be as realistic as possible. This realism is used to help the trier of fact observe the reproduction with very little explanation from the witness. In the area of importance, it is necessary to bring out the detail of the original. Remember, the model must be a fair and accurate representation. For example, in the case *Gerrard v. Porcheddu* (243 Ill. App. 562), a detailed model was a complete replica of the instrument used to ignite a house fire. Based on the fact that the model did not deviate from the original and it was used as illustration to a witness' testimony, the model was found to be properly admitted into evidence.

When components are not technically important within the model, then a simple design, called stylization, may be used. This form of construction gives an abstract and natural sensation to the model's component and, when placed within the model, helps determine the model's correctness in size and proportion. Juries throughout the United States have accepted stylization as long as the original general shape and color are maintained in the model. For example, blocks of wood can be used in a street scene to represent buildings. If the detail of the buildings is not important, then the blocks should look as close as possible to those buildings in shape and color. What is important is the location of the building in reference to other buildings or to the relationship of reference points within the overall model. In a previous mentioned court case, *People v. Speck*, wooden blocks were used to indicate the position of the victims within a townhouse.

Another example involving stylization concerns the court case *Beverly Hills Fire Litigation* (695 F 2d 207). The United States Court of Appeals, 6th Circuit, gave its opinion on this case involving, in part, a model of an interior wall containing an electric outlet. This outlet was purportedly to have started a fire at the Beverly Hills Supper Club in Southgate, Kentucky. Concerning the model, Circuit Judge Engel stated:

> The court did not err in permitting the employment of external portions of a model of the north wall. It was not found to be inaccurate, and it was used for demonstrative purposes by both plaintiffs and defendants. It further was not error for the trial judge to preclude the plaintiffs on rebuttal from eliciting proof that the interior construction of the model did not conform to the actual construction of the wall. Because the parties learned of the inaccuracies early in the trial, the judge never allowed the jury to view the interior of the model. It would not have aided the jury's consideration at all to have expanded the proofs further by developing this irrelevant information.

Once the decision has been made as to what the model is going to portray and the design of the model established, the next step is to determine the kind of material to be used in the construction. The choice of material will depend on established cost factors, the amount of man-hours dedicated to the project, the availability of the material, the type of model, what the model is going to illustrate, tools available, and the skills of the model maker. The important thing to remember is the texture of the material should look as close as possible to the original.

The model maker's first time handling new material should not be at the start of the model's construction. To gain proficiency in working with new material, practice with scrap pieces. Learn what tools are needed to cut the material. Work with different adhesives to find the best results for each type of material. Discover what type of paints will hold best on the material. In essence, the more experimentation performed on the material, the more the model maker will be aware of the limitations of materials and develop skills needed to work with the materials.

Another way to develop skills and increase knowledge of material is to observe how other models are constructed. Make notes when observing how other models are constructed. Go through a thought process as to why the model builder picked a particular material or how the material was constructed to look like the original. Determine if the overall concept of the model is compatible to what you believe the model is illustrating. If contrasting ideas exist make notes as to how you would make the model different.

MATERIALS FOR MODEL CONSTRUCTION

For the novice model builder there are many types of materials, tools, and construction techniques that can be utilized. The following materials are easy to work with, lightweight, and inexpensive.

Form Board

This material is constructed with a thin layer of styrofoam in the center with heavy colored paper or cardboard covering both sides making the foam board very rigid. The material is light weight and durable and is easy to cut using a straight edge and a razor knife. To assemble the boards together use hobby white glue. Purchased in art and office supply stores, foam board can be found in a variety of sizes and thicknesses. Another name for this material is foam core board.

Cardboard

Cardboard is an excellent material to use throughout the construction of the model. It is cheap and easy to work with. Cutting is done with a straight edge and a razor knife. The best adhesive to use on cardboard is rubber cement or small amounts of white glue. Another type of cardboard, called corrugated cardboard, is very sturdy and used for its rigidity. Found in various thicknesses, sizes, colors, and textures, it is used in commercial packaging and can be picked up for no charge. It is light in weight and easy to cut with a straight edge and a razor knife.

White Illustration Board

White illustration board is another popular material used for construction. Consisting of many plies of cotton fiber that makes it rigid, illustration board is easy to draw on and easy to cut. The color of the board is also white, allowing the edges to be colored or left plain. Illustration board is excellent for showing the different thicknesses between interior and exterior walls. Foam board is used as exterior walls and the illustration board in used for the interior walls. Another feature of illustration board is its pliability, enabling the material to be used when building contours need to be shaped.

Paper

Heavy white typewriter paper is good to use for elevation drawings of windows and doors. These exterior openings can be photocopied to reduce construction time when there is more than one of the same style and size. By using a photocopier with an enlarging/reducing capability, photocopies can be made in proportion to the model's scale. Art and office supply stores carry the paper in white and different colors and textures.

Wood

Wood is the second most used material after the paper products mentioned above. This material is excellent for its rigidity, dimensional accuracy, and ease of fabrication. One of the best features of wood is how abundant it can be found. The quality of the wood is determined by its hardness and direction of grain. Wood hardness range from the light balsa wood that can easily be cut with a knife to the hardwoods that come from the broad leaved trees that drop their leaves every fall. In between are the softwoods that come from the needle leaved evergreen trees. The softwoods, as a rule, are less expensive, easier to tool and more readily available than the hardwoods. The best wood to use in model building is called wood lumber. Purchased according to length and width, the optimum dimensions to work with are 1" x 2", 2" x 2", 2" x 4", 2" x 6", and 2" x 8". Wood lumber is excellent to utilize in stylization for structures.

Wood grain is the direction and alignment of the wood fibers that run up and down a tree and along its branches. The two types of grain to look for are the vertical grain, which runs parallel with the length of the lumber, and the flat grain, which has a marbled appearance. The type of grain depends on how the wood was cut at the mill. For a stronger piece of lumber, look for the vertical grain.

There are two basic cutting techniques used on wood. Cross cutting is a cut that is made across the lumber (across the grain), slicing the wood fiber. This is the easiest cut to make. The second type of cut is the rip cut, which cuts along with the grain and scrapes and tears the wood fibers. This action takes more work to accomplish.

Wood Products

Wood products consist of plywood, particle board and hardboard. This type of material is composed of wood byproducts and held together with assorted adhesives. These sheets are mainly used as the base foundation of the model.

Plywood

Plywood is manufactured from thin sheets of wood veneer that are glued together with the grain of each veneer running perpendicular to the layers just above and below. The standard size of plywood comes in a 4' x 8' sheet, and can be cut to any smaller dimension at the lumber yard. Thickness varies from 1/8" up to 1". Thickness between 1/2" to 5/8" is ideal due to its rigidity and weight.

Particle Board

Particle board is constructed from chips and particles of waste wood that are held together by synthetic resins. Due to its poor density, particle board lacks strength and is not a good choice as a base foundation. If using particle board, do not use a water base paint. The water will soak into the board and raise the grain. Most adhesives do not bond well with this material.

Hardboard

Hardboard or fiber board is produced by first reducing waste wood into fibers, then bonding the fibers under pressure with adhesives. Most adhesives do not bond well with this material.

Paper Mache

In model making, paper mache is used primarily to build landscapes in topography models. The concept is to take sheets of newspaper and tear them into strips. The width and the length will depend on the model size. The paper is then soaked in a wallpaper paste. After soaking the newspaper, it is applied to the surface (in model making usually a wire mesh) in several layers. In between the layers, add a coat of the wallpaper paste. Allow to completely dry before further work. The final product will dry hard and is then painted to the desired color.

Fasteners

The fastest way to join two pieces together is with fasteners. In model making, the three most used fasteners are adhesives, nails, and screws.

Adhesives

To keep the model pieces together and bonded to one another, adhesives known as white glue, contact cement, and super glue will be the most used by the model builder.

White glue consists of particles of synthetic resin suspended in water. This type of adhesive works best with porous and semiporous surfaces. The adhesion process takes hold when the water evaporates leaving the synthetic resin bonded as a clear film between the two surfaces. On paper products, white glue has a tendency (because of the water) to wrinkle the paper. A similar product of white glue, called carpenter's glue or woodworker's glue, works on the same principle with a stronger bond. The advantage of using white glue is its property to bond after several minutes of application. The time interval gives enough time to properly position the material in the desired location.

Contact cement is used when white glue cannot be used. This type of adhesive is used when evaporation cannot escape between the two materials. Contact cement is applied to the surface of each material. Once the cement partially evaporates (after several minutes), the surfaces are pressed together.

Super glue can be used on both nonporous and porous materials. This means that almost any type of material, paper, glass, plastic, metal, etc., can be bonded. Super glue works quickly in setting up and has a permanent hold. Be very careful when working with this type of glue. Do not get super glue on skin surfaces and then touch a material or bring the fingers into contact with one another. There will be an instant bonding that may require medical attention to remove.

Spray adhesive is used for paper and light cardboard. Spray adhesive creates a messy work area from overspray. To solve this problem, find a large cardboard box. Cut and remove one side of the box making a temporary spray booth. To use, spray a thin layer on both surfaces that will contact one another and let semidry. The disadvantage of using adhesive spray is that once contact is made of the material with adhesive there is not time to position in the desired placement.

Another type of adhesive used with limited success is the electric glue gun. Tubes of glue are placed in a form of a gun where heat is applied causing the glue to turn to a semiliquid. The heated glue is applied to the material, where it will cool, binding the two surfaces together. The glue gun is best used for the pin-point gluing of small items onto bigger surfaces. The adhesive on joints is not strong and could leave a stringy mess on the model.

Before any type of adhesive is applied to a material, the surface must be prepared to accept the adhesive. By following six steps, a proper bond will develop.

1. Clean the joint or surface of any foreign material.
2. With a fine sandpaper, rough up the surface to be glued.
3. Do not touch the surface with fingers as skin oils may interfere with the adhesive.
4. Apply a thin coat of adhesive in an even layer.
5. Bring the two surfaces together at the appropriate time given in the adhesive directions.
6. Let the proper time interval elapse for drying.

Nails and Screws

Nails are usually a finish nail called brads and are used to join thin pieces of wood together or two sensitive edges together. Finish nails are used when the head is not to show.

There are numerous screws that are available when working with wood. The three types of wood screws, in various sizes, that are used are:

Flat head –sits flush with the material's surface.

Oval head–partially recessed screws used when a attaching hardware.

Round head–sits on top of the surface and is used to hold thin material between the screw head and the surface.

Paints

The color of the model can become an important issue when duplicating the original. The defense attorney could have a field day with the model builder if the original color did not match that of the model. Therefore, paint becomes an issue in the fire scene model construction.

There are two basic type of paint. A water soluble base paint called latex that will dry in a flat, semi-gloss or high-gloss finish. Sometimes the word "acrylic" will precede latex. This means that a plastic resin has been added. Since water is the base of the paint, the cleaning agent is also water. Latex paints have a tendency to dry a lighter shade from the color shown when first applied.

The second type of paint has a vegetable oil base or a synthetic oil called alkyd. The oil base paint dries to a darker color after being applied. This type of paint needs special solvents to clean up with.

To apply paint to the surface of a material, spray painting with an air brush, aerosol can, or a brush can be utilized. Spray painting can be used for large areas where a uniform color is preferred. When spray painting, start from the top down on a vertical surface. Start the spray before the actual area to be painted, and continue in a straight uniform stroke. The distance of the spray nozzle to the material should be constant. Discontinue the spray after passing over the area to be painted. The next pass of the stroke should overlap at least one-half of the pass above. The distance of the nozzle, to the material surface determines the amount of paint being applied. The closer the nozzle the more paint is being deposited. Aerosol cans of spray paint take the place of spray guns for small to medium projects. Use the can approximately 12 to 16 inches from the object. There are two disadvantages to using aerosol cans. First, the nozzle is not adjustable for working in smaller areas like that of the air brush or spray gun. Second, different cans of the same color may produce slightly different shades.

Brush painting in large areas is uneven and can leave brush marks within the dried paint. Therefore, use the brush for small applications. Before undertaking the painting of the model, try the paint on a small piece of material that is going to be used. Check to make sure the color does not bleed and that it is the right shade when dried.

Another form of coloring is with felt-tip markers. Instead of paints, the markers use either a solvent base ink (permanent ink) or a water base ink (watercolors). The solvent base ink is best used for model building. Sometimes the novice, when using a felt tip marker to cover large areas, has a tendency to allow it to appear like Venetian blinds from the overlapping strokes and ink bleeding. The colored markers are excellent to use in small areas or for colored lines and should become a part of any model building tool kit.

Tools

Whether a beginner or an advanced model builder, a basic tool kit is needed. All tools should have a high quality to them. Tools must be kept clean and must be stored properly. When you are finished using a tool, get in the habit of putting it back in its proper place.

The basic list:

cutting surface	utility knife
squeeze type clothes pins	Xacto® knife
various adhesive tape	pencils
straight pins	pens
single edge razor blades	rubber bands
scissors	small finish nails
6" steel ruler	tweezers
18" steel ruler	assorted line weight
architecture ruler	assorted line weight
45 degree angle	90 degree angle
electric glue gun	T square
assorted felt tip markers	assorted types of adhesive

The cutting tool most often used will be the razor knife. To properly cut a piece of material:

1. Have a sturdy cutting surface that will stand up to the constant contact of the knife.
2. Place the straight edge on the desired mark to be cut.
3. Place the point of the knife at the beginning of the cut. Press firm ly down penetrating the material and draw the knife along the straight edge. Do not worry if the cut does not go through the material. Overpressure on the knife will bend the blade and force it away from the straight edge. Just continue the process of cutting described above until the blade is completely through. If the cut is not completely through and an attempt is made to separate the material, a tear may develop forming ragged edges.

For the advanced model builder, power tools will expand the list of materials for construction. New and innovative techniques will develop, adding to the model builder's expertise. Once proficient in their application, power tools will be able to decrease production time.

Work Area

To avoid problems when constructing the model, the work area should be divided into three-separate sections: the drawing/ sketching, the cutting, and the assembly areas.

Drawing/Sketching

The surface can be a desk or drawing board. When using the three section work area, this section should be off to one side because of its limited use. When space is limited, this area can also be used as the assembly section and is the center of the work area.

Cutting

This section should be constructed of a material durable enough to withstand continuous cutting. The section is located within easy reach. The majority of the model's construction will be between the cutting and assembly section.

Assembly

This section is the center of activity when in construction. The table should be strong enough to withstand the weight of the model and the pushing and pounding of the model builder's work.

Each section is designed and organized for a specific task. A clean and neat area with plenty of room will alleviate unnecessary mishaps and delays that might ruin the model. The total work area should be well lit, with plenty of ventilation while using paints and adhesives. Electrical outlets may be utilized by the advanced model builder when using mechanical electrical tools.

BUILDING THE MODEL IN TWO PARTS

The fire scene model, whether as an architectural or engineering model, is constructed in two parts, the base and the site elements.

The Base

The strength of a structure is derived, in part, from the ability of the foundation to support the structure. This concept also applies to the model. Without a good foundation or base plate, the model will be constructed with flaws that could damage the model beyond its capability as a legal model. Every type of model will need some type of foundation to support the exhibit and to prohibit the model from warping and bending.

The base plate for the architectural model should be a simple, rigid piece of plywood. The size of the base plate will depend on the size of the model and whether the model is standing by itself or integrated into a neighborhood site plan. The most popular shape to use is a rectangle for its ease, dimensions, cutting, and transportation. The underside of the base should contain at least one rubber pad on each corner. This will protect the surface area that the model will be resting on. To stop the base from warping, a 1" x 2" wooden furring strip is fastened to each of the four sides. When features become important beyond the building lines, plan to extend the base so that the physical elements can be observed. This type of model will show adjacent buildings, landscape, walks, roads, topography, and evidence.

On the base, the first step in the construction of the architectural model containing a building, floor plan, neighborhood or a diorama is to find the center line for both the length and the width and mark it. At the center line, the model's exterior dimensions are centered so the model will sit on the base plate with equal distance on each side. Once the exterior walls are located on the base plate then, if needed, the floor plan can be added. With the exterior and interior walls marked, a type of template is developed to use as a guide for the erection of the walls.

The second part of building the model is the placement of the site elements. The site elements will depend on the type of model being constructed.

One of the most used site elements will be the interior and/or exterior walls of the model. Before making a cutting of the exterior walls, determine what type of joint will be used to bring the walls together. Through preplanning, the type of joint will dictate the overall measurement of the length of the wall.

Joints

The joint brings two pieces of material together with an area containing the adhesive that will hold the materials in place. The three most used joints are the miter, butt and the butt joint with a cover.

Miter Joint–Figure 11-11

The miter joint is used when the material has a noticeable thickness that will affect the overall measurements and aesthetic look of the corner. The wall's overall length is measured and marked on the outside section. From this mark, a 45-degree angle towards the interior is cut from floor to the ceiling. When joined to another wall with the same type of cut, a 90-degree angle is formed leaving a small seam at the corner. This type of joint is the hardest and most time-consuming joint to construct. This joint is used mainly for wood or plastic materials.

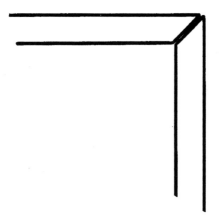

Figure 11-11. A miter joint.

Butt Joint—Figure 11-12

The butt joint is a straight cut of the material. The butt of one wall is joined to the interior surface edge of the second wall. In this type of construction some of the walls need to be compensated for the material's thickness. To avoid the problem of figuring out which wall needs measurement correction, a simple rule can be applied. Measure all east and west walls at their full length. Next, measure all north and south walls at their full length minus the thickness of the east and west walls. This type of joint is the simplest and easiest to construct and can be used with any type of material.

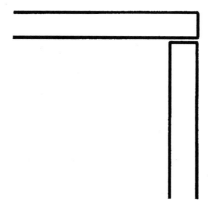

Figure 11-12. A butt joint.

Butt Joint with a Cover—Figure 11-13

Like the butt joint, the material has a straight cut from floor to ceiling. Measure all east and west walls at their full length. Also, measure the full length of the north and south walls. From the edge, the thickness of the north and south walls are marked on the inside surface of the east and west walls. With a straight edge, Lightly cut into the material with a sharp knife or razor blade on the mark from floor to ceiling. The object of the cut is not to penetrate the opposite surface. After the cut, remove the material from the edge to the cut line, leaving the outside cover intact. The joint is best used on foam board. The effect is the same as the miter joint with less work.

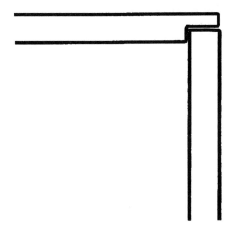

Figure 11-13. Butt joint with a cover.

To give further strength to the joint a support from the same type of material used for the walls can be utilized (Figure 11-14). Cut a small piece of the material the same dimension as that of the floor to ceiling and wide enough to adequately cover each side of the joint. On the two sides where the pieces will come into contact with the walls, add a thin bead of glue, then place it in the interior corner.

Figure 11-14. Strengthening a joint.

To strengthen a splice (Figure 11-15), adhere a rigid piece of material to the side of the model not in view. The width of the material should be wide enough for good coverage of both sides of the seam. The length should be the same as the splice.

Figure 11-15. Strengthening a splice

Once the type of joint has been determined, begin construction by making and cutting out the first exterior wall of your choice on a flat surface. When measuring the length of the material for either the exterior or interior walls, remember to allow for the thickness of the material. (For an example, see figure 11-16). Next, cut out the basic components such as windows and doors. To save time, photocopy the type of windows needed from an architectural book. The exact size may have to be reduced or enlarged to correspond proportionately to the model. Construct the next exterior wall in the same manner as the first one. Before applying a thin layer of adhesive, test each wall for accurate length and height. If satisfied, glue the two walls together in the shape of an L and onto the base plate. Each of the remaining exterior walls should be constructed, tested in the same format, and separately glued to its proper location on the base plate. If a floor plan is being incorporated into the model, the interior walls can now be constructed once the interior walls are in place.

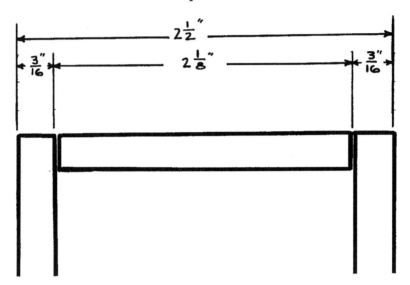

Figure 11-16. How to measure for an exterior/interior wall.

Other Types of Site Elements

Once the base plate or the topography is completed, the model evolves in detail by adding surfaces and vertical elements. Vertical detail should always correspond to the scale of the model or be proportionate to one another. Small details may be omitted as long as the items are not important issues in the case. For example, if a structure has rain gutters, the model may omit the item. Site detail can be divided into two groups. The first group adds detail to the building, such as trees, urban and city features, water and paved surfaces. The second group, directly connected to the structure, is the vertical detail consisting of widows, doors, and staircases and the horizontal details which are beams, handrails, and balconies. When floor plans and dioramas are used, interior furnishings are included as detail connected to the building.

Detailing can include any type of material that resembles the object being reproduced. It should be simple and easily identified. Through experimentation with new materials, skills, and techniques, many different procedures can be produced with the same result. Record keeping of each new experiment can be used for quick reference.

Detailing begins at the ground plane and works up. At the ground plane add material that represents water surfaces, paving surfaces, and ground cover plant life.

Water surface detail, in its simplest form, can be blue paint or a felt tip marker shading the area directly on the model base. Active water can be symbolized by using typewriter correction white out fluid to add the texture of white caps. Advanced techniques can consist of using aluminum foil covered by a blue color acetate or a fluid epoxy especially made to resemble water surfaces. These epoxies can be found in hobby and arts and crafts shops.

Paving surface, like water surfaces, in its simplest form can be black paint or felt tip marker for asphalt. Rolled tar paper is another form used to resemble asphalt. Grey cardboard can be used to resemble concrete driveways and sidewalks. Sandpaper, based on the texture and color of the paper, can represent many different types of materials. When using sandpaper, use rubber cement instead of white glue. White glue has the tendency to warp or wrinkle the paper. Flagstone, brick, and other types of pavers can be drawn directly on the base. The use of photographs of the original item can be used and glued to the base. The photograph must be proportional to the model.

Plant life is divided into two groups. Ground level plants and vertical plants. Ground level plants are applied first and represent grass and plants under two feet in height. The simplest form of ground cover can be applied as green paint or felt tip marker. Also, artificial grass can be purchased from model train hobby shops. Plants under two feet can be represented by lichens found in the woods; sponges, especially loofah; and model plants bought commercially.

To properly show the correct proportional relationship of buildings, vertical plants such as trees and shrubs become important details. For instance, if a tree is too low, the building will look too tall. If the tree is too large, the building will look too small. Stylization in the form of dried plants, flower heads, twigs, and branches have a natural appearance of some vertical plants. Store-bought model railroad plants and trees have a more natural resemblance to the originals. Whatever form of detail used, first study the plants around the fire scene. Note how the trees and shrubs look in their natural state. Once the detail form has been determined, do not mix different types of materials to resemble vegetation.

The purpose of site detail, such as utility poles and street signs, is to give the viewer an idea of the model's scale or to interpret the ratio of sizes. Site detail can be constructed when using a large scale or to save time in man hours kits may be purchased at hobby shops in the Lionel, HO, N, O, scaled kits.

Details connected to the building can be drawn, cut out, or photographed, and pasted onto the elevation. Again, to save time, commercially made prefabricated plastic kits from the model railroad and doll house hobby industry may be used. Small details such as handrails, balconies, and fences can be constructed from solder wire, the remains from plastic trees of car model kits and small pieces of wood.

LEGEND

Any model that is constructed should contain a legend to indicate the following information:
- Address of fire scene or purpose of model
- Date and time of the fire
- Name of model building
- Type of scale
- Direction (indicating north)

Finally, the opposing attorney will seldom object to the use of a model for fear of not wanting the jury to think they have something to hide. A good opposing attorney will let the model be introduced as evidence, then use the model to advantage during his or her presentation of the case. Opposition will occur if the model is not a true and proportionate representation of the original. Some attorneys have even objected that a model was not the original and may give a false impression to the trier of fact. For these reasons, it is imperative that if time and money are going to be spent on constructing a model, then the builder and the fire investigator must be sure the model is a complete representation of the original.

With minimum knowledge of specialized skills, the use of readily available materials and practice, a fire investigator can construct a true and accurate model of the fire scene for courtroom presentation. The model is used as an aid to the witness' testimony and enables the trier of fact to realistically imagine how the fire scene appeared before the fire, what the scene looked like during the fire and how the fire scene was investigated after the fire. If the appearance is misleading in scale, proportion, texture, or style, the model may be found inadmissible for trial use.

Chapter 12

COMPUTERS AND THE FIRE SCENE INVESTIGATION

The newest form of demonstrative evidence to revolutionize courtroom presentation embraces the use of the computer. With computer systems becoming smaller and more affordable, the legal profession is turning to the computer as an avenue to help prove their cases. In both civil and criminal litigations, a trend is beginning to emerge that uses the computer to create images on a visual monitor. This technology, called computer animation, can recreate an accident, difficult crime scene, or fire scene in a series of still images that are shown in rapid succession to produce the feeling of motion. This movement aids the trier of fact to better understand how the dramatic events developed. Computer animation can take the viewers in the courtroom to the event in question and observe what occurred from any perspective–front, back, overhead, below looking up, any important angle or in a cut-a-way, any machine, vehicle or structure, in an instant. The viewer can even be placed in the position of an eyewitness and observe exactly the same details as the eyewitness. Computer animation can also be utilized when the events are too expensive, hazardous, or physically impossible to recreate. This makes computer animation an excellent form of demonstrative evidence when used to depict a fire scene.

At one time, the only way to show a re-creation was through the use of motion pictures and/or video. Actors had to be used, scenes constructed to resemble the original, each camera could only shoot the re-creation at one point of view, and the re-creation had to be filmed as close as possible to the original conditions documented at the time of the event. This type of video re-creation was expensive and many times not completely accurate or realistic.

249

Using completely different techniques, the computer can use software to produce 3-D animated programs from data taken from the scene. Several years ago, computer animation could only be performed on mainframe computers, employing many man hours to complete the project with a very high cost factor. Today, with microcomputers, PC's (personal computers), microprocessors, and advanced technology associated with them, programs can be easily learned at home or in the office at a reasonable price.

To show the difference between a video and computer animation, let us examine a theoretical crime scene involving a homicide and fire to cover up the murder.

At 2:30 AM, a fire department responded to a single family residence for a structure fire. The initial call came from the next door neighbor who was awakened by an orange glow coming into the bedroom window. Looking for the source, the neighbor observed flames inside the window opening of the residence next door. While suppressing the fire, first-in fire personnel found a deceased, fully clothed, male victim, face down next to the bed. During the death investigation, the victim was rolled over where a pool of blood was observed. The victim sustained severe thermal injuries to the head and side of the body closest to the bed. The victim was removed and the fire investigation was conducted. The point of origin was determined to be on the bed away from any type of electrical wiring or appliance, ruling out electric as a fire cause. Evidence within the dwelling indicated the victim was a heavy smoker. Food was found in the kitchen in a pot located on the stove. The kitchen table contained several empty beer bottles. Evidence of forcible entry was established to have been from the firefighters gaining entry. No other signs of forcible entry or abnormal entry were detected. The preliminary determination was that the victim fell asleep on the bed while smoking a cigarette. However, because the cause of death had not been determined, and the possibility existed of a crime scene, the scene was secured until the conclusion of the medical examiner's autopsy.

The medical examiner conducted a complete set of x-rays of the fire victim and discovered several 9 mm bullets in the head and chest cavity. Each of the six wounds was documented as to location, entry angle, penetration, and organ damage. The autopsy further showed that the victim was not alive during the fire. The medical examiner determined the cause of death to be multiple gunshots, three of which were life-threatening. After receiving the medical examiner's report, the fire investigators went back to the fire scene with a search warrant. During the daylight fire scene investigation, two bullet holes were discovered in the upper areas of the exterior wall. The discovery was made when one of the investigators observed sunlight coming through the holes. The bullet holes were not detected during the nighttime investigation due to the darkness and the carbon deposits of the fire surrounding the holes. Following the trajectory from the holes, the investigators found a projectile within the underside of the eaves. All observations and evidence taken were photographed and documented.

Several days later, the wife was arrested for the murder of her husband and first degree arson to cover the crime. The wife claimed that she left the residence after her husband came home intoxicated and attacked her. The fire must have occurred from a cigarette since the victim was smoking. On the other hand, the prosecutor, claimed the wife, in a rage of revenge, waited for her husband to fall asleep, then shot him once in bed. Startled by the gunfire and wound, the victim rose from his bed and tried to defend himself as the wife continued to shoot. The victim fell dead next to the bed. Hoping to destroy evidence and conceal the gunshot wounds the wife set the bed on fire by pouring and igniting charcoal lighter fluid.

To assist in the presentation of the state's case, the prosecutor felt it necessary for the jury to visually see how the events developed. By using either a video or a computer animated reenactment of the scene, three goals could be accomplished. First, the jury would get a better understanding of the case by displaying a reconstruction of the scene. Second, the visual record would help illustrate the expert's testimony of how the murder and fire occurred. Third, the video or animated scene could be used as a source to impeach the defendant.

One of the problems confronting the state's attorney was deciding what form of demonstrative evidence to use—video or computer animation.

To produce a video, a large expense account would be required for materials and the hiring of contractors to reconstruct the crime scene. Also, finance would be needed for actors and a filming crew. To add to the expenditure predicament, technical difficulties exist as to how the trajectory of the ballistics would be demonstrated as well as the progression of the fire. A part of the reenactment would be rendered to an animated cartoon, creating flaws in accuracy and true representation of the actual crime scene. In all probability the judge would rule the video inadmissible as demonstrative evidence.

The second form of demonstrative evidence considered by the prosecution was the computer animated reenactment. To the state's advantage, the animated production through information gathered at the scene would show a life-like scenario: the location of the victim and the defendant, before, during, and after the shooting; the path of the bullets as they left the weapon and at what sequence they hit their target. The scene could be duplicated over and over again from a different perspective, show the event at actual speed or slowed down for clarity, using a close-up (without distortion) to focus attention on a particular issue, all in an instant by giving the command to the computer. Last, the computer animation would support the expert's testimony as to how the fire was willfully set by using an accelerant and disproving the defense's case that the fire started by a cigarette.

Legally, the two separate classifications of computer-animated evidence are computer animation and computer simulation. In appearance, both look similar on the monitor or television screen. However, the distinction lies in the definition of how each individual classification programs the movement observed. To comprehend the difference between the animated and the simulation a discussion on each follows.

COMPUTER ANIMATION

The development of computer animation for litigation began in the late 1970s and early 1980s. Used mainly to show movement and perspective, computer animation replaced the canvas or drawing board in the courtroom and used the computer and its monitor for the graphic details of an incident. The early stages of computer animation involved stick figures or crude images of objects in a two dimensional plane. Technology advancement generated clearer images that were more realistic, making it an excellent tool for creating reenactments of accident scenes.

The animated imagery observed on the monitor is only as good as the data taken from the scene. The investigator must make accurate measurements, precise observations and good photographs so the animator can program the computer for the proper characteristics of size, shape, color, and path of travel of objects. The animation makes no assumptions by itself and relies entirely on the expert's data.

Computer animation may be found inadmissible on the grounds that the animation is being unfairly prejudicial. This idea is based on the fact the input data is supplied by the programmer with the illustration used to mislead the trier of fact. For example, the expert witnesses for the prosecution may classify a fire as incendiary and establish the area of origin to be a pour pattern on the floor. The defense expert witnesses may disagree and believe the fire to be accidental with the areas of origin at an electrical outlet on a wall. To further draw attention to their justification, the defense creates a computer animation to interpret the accidental fire based on the information gathered by their experts. The prosecution may argue that the data used to illustrate the sequence of events may have been influenced by improper methods, incomplete information, or false information intentionally given. Second, another argument may content that the illustration is merely a cartoon and combined with the first argument of being prejudicial, the computer animation does not have the ability to be a true and accurate representation of what occurred. Like other forms of demonstrative evidence, the final outcome for admissibility of the evidence is up to the discretion of the judge.

COMPUTER SIMULATION

In the early 1990s, computer technology produced new super computers and microprocessors that can store vast amounts of information and perform calculations at a faster speed. Through this technology, the illustration observed on the monitor has advanced to a three-dimensional image. Computer simulation is based on two principles. The first principle, similar to computer animation, is derived from mathematical calculations taken from the scene. The second principle, which separates computer simulation from computer animation, involves the input of data applicable to movement and the physical laws that influence its motion. Through the combination of mathematical calculations and laws of physics, a scene can be analyzed, developed as a reenactment in 3-D, and by filling in missing data arrive at a conclusion based on assumptions. In other words, with the information given to the computer, it can reason a solution to a problem. In an article written by Louis A. Lehr, Jr., "Admissibility of Computer Simulation" (For the Defense, July 1990), Mr. Lehr gives a clear and simple definition of computer simulation:

> Computer simulation is the result of a computer analysis of a model which represents the actual structure and is subject to the relevant physical laws governing its motion. The programmer does not dictate the outcome of the analysis; this is dependent upon the model itself and the input data.

Computer simulation can contribute to the litigation of a case during the stages of strategical planning. First, the simulation should be produced as soon as possible. This will provide the attorney with an advantage in comprehending the expert's testimony through a visual illustration. Once the visual reenactment becomes clear to the attorney, other witnesses can view the simulation to review testimony, develop important questions, and address critical issues. A game plan can then develop as to how the attorney will present his or her case, answer "what if" questions, help generate arguments for the opposing attorney's attack, and to use computer simulation as rebuttal evidence to disprove the adverse party's evidence. Second, after the simulation is completed, new evidence may be obtained and programmed. The computer will then recalculate all the information and display a new image that may strengthen or weaken the demonstrative evidence in the case. In either circumstance a new approach of how to utilize the evidence or how to present the case may be necessary.

Another valuable asset of computer simulation in litigation is to display the graphic image during trial. Like the animated picture, the simulation is shown to the trier of fact as an actual reproduction of what occurred at the scene. The movement can be presented in real time, slowed down to capture fast movement, or advanced to accelerate slow movement. The image can be placed in a still frame mode to stop the action and focus on important details. The simulation can appear from different perspectives through any combination of overhead, side, front, or back views. By cutting away the outside exterior, this form of simulation called "cut away" will reveal the interior working parts. In each of the ways to present simulation, the important thought to remember is how the trier of fact will perceive the simulated evidence, and, if it is the best and most accurate means form to demonstrate your position of how the incident developed.

When to Use Computer Simulation

Not every court case will use computer simulation or computer animation as demonstrative evidence. A number of factors should be considered before implementing an event through computer simulation.
1. Certainly the most important criteria would be the amount of property loss, number of injuries, or loss of life.
2. A simulation is used when a critical scene or the understanding of an object cannot be visualized through testimony.
3. When the scene or object is too dangerous or the cost too expensive to reproduce the original.
4. When movement cannot be shown clearly through other forms of demonstrative evidence.
5. A computer simulation will help solve conflicting eye witness accounts or disprove the view of the opposing witness.
6. When the opposing expert witness has a different opinion as to how the event occurred.

ADMISSIBILITY OF COMPUTER ANIMATION AND SIMULATION

Throughout the history of courtroom drama, the courts have allowed witnesses to reenact a particular event, reenact an event through the use of film video, and have allowed the testimony of expert witnesses based on staged experimental reenactments. Computer simulation is another interpretation of the reenactment using the latest technology.

One of the earliest court cases to use computer simulation was *Perma Research and Development v. Singer Company* (542 F 2nd 111 [2d Cir], 429 US 987) in 1976. In this civil litigation, the plaintiff, Perma, filed suit against Singer for a breach of contract. As part of Perma's evidence, a computer simulation was used to corroborate their expert's testimony. On appeal, Singer cited that computer animation should be inadmissible on the grounds that there was no information passed to the lawyers representing Singer to verify the calculations of the simulation. The appellate court disagreed and stated that it was the trial judge's discretion in allowing the expert to testify.

The significance of this decision rested not in the opinion of the court, but in the dissenting opinion of Circuit Judge Van Graafeiland, who responded that careful consideration must be made when accepting computer technology as evidence. Judge Van Graafeiland stated:

> As courts are drawn willy-nilly into the magic world of computerization it is of utmost importance that appropriate standards be set for the introduction of computerized evidence.

(Judge Van Graafeiland then quoted Jerome Roberts, A Practitioner's Primer on Computer-Generated Evidence. 41 U. Chi. L. Rev. 254, 255-256, [1974])

> Although the computer has tremendous potential for improving our system of justice by generating more meaningful evidence than was previously available, it presents a real danger of being the vehicle of introducing erroneous, misleading, or unreliable evidence. The possibility of an undetected error in computer generated evidence is a function of many factors: the underlying data may be hearsay; errors may be introduced in anyone of several stages of processing; the computer might be erroneously programmed, programmed to permit an error to go undetected, or programmed to introduce error into the data; and the computer may inaccurately display the data or display it in a biased manner. Because of the complexities of examining the creation of computer generated evidence and the deceptively neat package in which the computer can display its work product, courts and practitioners must exercise more care with computer generated evidence than with evidence generated by more traditional means.

Being a fairly new form of demonstrative evidence many jurisdictions are just beginning to deal with computer simulations. The outstanding quality of computer simulation, using scientific principles and calculations to arrive at a conclusion, may be considered as evidence with probate value, thereby creating a higher standard of admissibility. Not knowing exactly how to admit such evidence, many jurisdictions require that a pretrial hearing be held to determine if the computer simulation will meet the test of admissibility for scientific evidence known as the Frye Test and to others as the Frye Standard.

In 1923, the Federal Court of Appeals of the District of Columbia gave an opinion in the case *Frye v. United States* (293 F. 1013, 1014.) In this case, the defendant wanted an expert witness to testify on the results of a deception test he had taken. This was denied. The court felt that since this type of deception test had "not yet gained such standing and scientific recognition," the expert testimony could not be used. In his opinion, Associate Justice Van Orsdel stated:

> Just when a scientific principle or discovery crosses the line between the exper-
> imental and demonstrable stages is difficult to define. Somewhere in the twi-
> light zone the evidential force of the principle must be recognized, and while
> courts will go a long way in admitting expert testimony deduced from a well
> recognized scientific principle or discovery, the thing from which the de-
> duction is made must be sufficiently established to have gained general accep-
> tance in the particular field in which it belongs.

Several months after the Perma Decision in New York, the Supreme Judicial Court of Massachusetts, in the case *Shaeffer v. General Motors Corporation* (372 Mass. 171, 360 N.E. 2nd 1062 [1977), stated that in order for a computer simulation to be used in court, the test for admission of scientific theory must apply. Even though the Frye Test was not distinguished by name in the Shaeffer Decision, it certainly was implicated through the interpretation. Justice Quirico, in his opinion, stated:

> Our concern is not with the precision of electronic calculations, but with the
> accuracy and completeness of the initial data and equations which are used as
> ingredients of the computer program. More generally, we feel that the standard
> for admissibility of scientific tests may not have been met in this instance. That
> standard was clearly enunciated in *Commonwealth v. Fatalo*, 346 Mass. 266, 269,
> 191 N.E. 2d 479, 481 (1963): "Judicial acceptance of a scientific theory or
> instrument can occur only when it follows a general acceptance by the
> community of scientists involved.

In an Arizona Court of Appeals decision, *Starr v. Campos* (134 Ariz. 254, 655 P.2d 794 [1982]), the Court's opinion was that a computer simulation of a fatal vehicle accident was improperly admitted in trial. The Appellate Court based this decision on the fact that the trial court failed to conduct the Frye Standard on the simulation. Judge Birdsall, in his opinion, explained how the Court felt about using the Frye Standard when he stated:

> Nor is it necessary that scientists agree that the results of the procedure will
> always be correct. If that were the standard, for example, the courts would cer-
> tainly be unable to allow the testimony of psychiatrists. The scientists need
> only to agree that the procedure has a sound scientific basis and is capable of
> producing a result that can be used, with awareness of any limitation for sci-
> entific purposes.

The Court also felt that if there was a retrial, the Frye Standard should apply to the computer simulation. Judge Birdsall stated in the opinion of the Court:

> If this evidence is offered in a second trial, therefore, the court is directed to apply the Frye Standard and determine specifically, in the absence of the jury, whether the procedure used to obtain that evidence is generally accepted among scientists in relevant fields, including accident reconstruction and automotive engineering. In making this determination the court may take judicial notice of the ability of a properly programmed computer to perform mathematical computation and of the general acceptance of the underlying principle of the method, the law of conservation of linear momentum. It will only be necessary to determine whether those of sufficient training and experience to judge are in general agreement that the program properly applies that principle (and any others it may involve) to automobile collisions.

One of the first criminal court cases to deal with a computer animation/simulation was *People v. McHugh* (476 NYS 2nd 721 N.Y. Sup. Ct. 1984). As part of the defense, an expert was used to prepare his version of the fatal vehicle accident through a computer-animated illustration. The prosecutor asked for a pretrial hearing based on the Frye Test as to "whether the computer program and the incorporated scientific and mathematical formulas, techniques, and processes underlying the program are generally accepted as accurate and reliable by the scientific community." The court felt that the Frye Test did not apply in this case since it was being used as demonstrative evidence, an aid to the testimony of the witness. Justice John B. Collins stated:

> The evidence sought to be introduced here is more akin to a chart or diagram than a scientific device. Whether a diagram is hand drawn or mechanically drawn by means of a computer is of no importance.

Realizing the use of a computer reenactment in trial was breaking into a new litigation frontier as demonstrative evidence, Justice Collins stated:

> While this appears to be the first time such a graphic computer presentation has been offered at a criminal trial, every new development is eligible for a first day in court.

A computer is not a gimmick and the court should not be shy about its use, when proper. Computers are simply mechanical tools–receiving information and acting on instructions at lightning speed. When the results are useful, they should be accepted; when confusing, they should be rejected.

Lastly, since the case appeared to be the first of its kind. Justice Collins established three rules of admissibility for the presentation of a computer-animated image as demonstrative evidence.

What is important is that the presentation be relevant to a possible defense, that it fairly and accurately reflect the oral testimony offered and that it be an aid to the jury's understanding of the issue.

To successfully use a computer simulation under the Frye Standard or Test the attorney must:
1. Irrefutably establish the accuracy of the simulaion.
2. Establish the credibility of the expert witness.
3. Demonstrate the simulation is based on theories long recognized under the laws of physics.

General Acceptance or the Frye Standard has been criticized in many courts as being too vague, not defining who is the relevant scientific community and that its functions are keeping valuable, reliable evidence from the trier of fact. A movement has begun to reject the Frye Test that has found its way into a United States Supreme Court Decision, *Dauber v. Merrell Dow Pharmaceutical* (113 S.Ct.) in 1993. This case involved a federal lawsuit in which a drug allegedly caused birth defects when ingested by two mothers during their pregnancy. The main issue in the case centered around the different scientific expert witnesses and whether or not the Frye Test of General acceptance applied. Aware of the many court cases involving the matter, the United States Supreme Court felt some type of standard should be determined. The Court stated.

We granted certiorari...in light of sharp divisions among the courts regarding the proper standard for the admission of expert testimony.

The first part of the opinion, delivered by Justice J. Blackmun, dealt with the origin and concept of the Frye Test. Unanimously, the Supreme Court stated that when the Federal Rules of Evidence was legislatively enacted, Rule 702 that governs expert testimony replaced the Frye Test. Rule 702 states:

If scientific, technical or other specialized knowledge will assist the trier of fact to understand the evidence or to determine a fact in issue, a witness qualified as an expert by knowledge, skill, experience, training, education, may testify thereto in the form of an opinion or otherwise.

The Court observed that the "general acceptance" standard set forth in Frye was not included in Rule 702 and therefore no longer applied. Justice Blackmun stated:

The drafting history makes no mention of Frye, and a rigid "general acceptance" requirement would be at odds with the "liberal thrust" of the Federal Rules and their "general approach of relaxing the traditional barriers to "opinion' testimony"... Given the Rule's permissive backdrop and their inclusion of a specific rule on expert testimony that does not mention "general acceptance" the assertion that the Rules somehow assimilated Frye is unconvincing. Frye made "general acceptance" the exclusive test for admitting expert scientific testimony. That austere standard, absent from and incompatible with the Federal Rules of Evidence, should not be applied in federal trials.

Justice Blackmun went further and stated that they (the Supreme Court) had confidence in trial judges to determine:

Whether the expert is proposing to testify to 1) scientific knowledge that 2) will assist the trier of fact to understand or determine a fact in issue...the Rules of Evidence–especially Rule 702, does assign to the trial judge the task of ensuring that an expert's testimony both rests on reliable foundation and is relevant to the task at hand.

It is advisable to consult your attorney when using computer animation or simulation to determine if the Frye Test or the Federal Rules of Evidence are applicable.

The basic rules for computer simulation to be admissible in trial are similar to the rules for all demonstrative evidence.

1. They must be relevant to an issue in the case.
2. Be authenticate by a witness with personal knowledge of the event or scene.
3. Must fairly and accurately reflect the testimony as to how the event occurred. This is very important when using other forms of demonstrative evidence. All details must match one another. For example, a couch in a sketch of a living room must also be shown in the photographs, video and computer animation/simulation. Also, the expert's data must be accurate and verified in court.

4. Good credentials are needed to show the knowledge, skill and training of the computer programmer.
5. The simulation is an aid for the trier of fact to understand the issues in the case.

THE FUTURE

The way we live today has been, and will continue to be, influenced by computer technology. As we approach the beginning of the 21st century, all crime scene investigators will see computer technology make a presence in the way an investigation is conducted and in the way evidence is presented at trial. To understand this impact, let's look at some recent advancements being used and what can be expected in the near future.

Fire Models

Fire models can be categorized into two general classifications: the pictorial fire model and the analytical fire model. Both are used to reconstruct how the fire scene appeared before the ignition sequence, during the fire's incipient stages, the fire's growth, effects of fire suppression activities, and how the fire scene appeared at the beginning of the fire investigation.

Pictorial Fire Models

Pictorial fire models began as computer models and as an aid to engineers, architects, and designers and are used to help develop and construct buildings. Through the use of a desktop computer, CAD (computer aided design) software, and a video deck/editing recorder, a computer image can be created in either a two-or-three dimensional display used as an architectural walk through of the interior and exterior of the planned building. There are two primary functions for using pictorial computer models. One, it is used as a time saving device that can create lines or objects determined by the designer. These objects and lines can be made by using a light pen, mouse, or graphic table. Second, the computer model can be programmed to show the image from any perspective view, and change the angle or direction through a simple command to the computer.

CAD is a software package designed to substitute work normally done by hand. There are many advantages to using CAD software.

1. The computer can automatically calculate dimensions.
2. It can easily add or delete changes.
3. Repetitive structural symbols can be added quickly.
4. The image can be enlarged to obtain a close-up view of a particular part or it can be shrunk to obtain an overall view.
5. The information can be stored indefinitely for future reference or reproduction.

Analytical fire models were originated to help design and analyze fire protection hardware. This type of model uses mathematical calculations to study important facts in a particular issue. Issues such as flame spread, heat buildup, smoke generation, and energy release rates assist engineers to determined the proper use of fire protection equipment. Taken a step further, these mathematical calculations can also be used as a tool to document a particular event in fire scene investigations and courtroom presentation.

The main objectives for using a fire model as part of the fire investigation may include the pictorial or analytical fire model or a combination of the two, to display one or all of the following:

1. Simulate the growth of the fire.
2. Determine the amount of smoke generated.
3. Produce a ratio of temperature to a timeframe to indicate heat buildup.

Other objectives may involve fire fatalities, product liability, fire protection failure, and flashover occurrence. The list is not limited to what had just been mentioned. As stated earlier, a solution to a particular issue concerning the fire or events that occurred during the fire may be explained through the fire model.

Like any other computer simulation, a fire model is only as good as the information it is given. Therefore, it is essential that accurate descriptions and dimensions of physical evidence from the scene be recorded. The more information obtained the better the simulation. Information should include height of the ceiling, overall dimensions of the rooms and the complete structure (this will give the square footage and volume of the area). dimensions of windows and doors, location of combustible materials including furniture and any other information that the investigator believes to be important should be recorded. The fire model can be used to help the trier of fact understand the witness' testimony, or it can be used to dispute the opposing side's view of how the fire started. For example, the state's expert witness has testified that multiple origins were found on the fire scene. Also, there were indications that an accelerant was used. The defense attorney disputes the state's claim, stating that the fire was accidental and that the cause of the multiple origins was due to a flashover. Through the use of the analytical fire model's temperature/time frame ratio, the state prosecutor can establish that the fire burned hotter and faster (due to the use of an accelerant) than what would be expected in the accidental fire and flashover claimed by the defense.

In 1987, the Federal Bureau of Alcohol, Tobacco and Firearms (ATF) began using computer fire models to help reconstruct large fire scenes. The first fire model used by ATF involved the New Year's Eve fire at the Dupont Plaza Hotel in Puerto Rico. This model was so successful that ATF continues to use fire models to document fire scenes. ATF feels so strongly about the use of computer fire models the following statement can be found in their CFI (Certified Fire Investigator) Course Handbook.

> In particular, the analysis of the Dupont Plaza fire by H. Nelson showed how a systematic application of fire formulas could explain the process of the fire growth and determine a consistent time line for the fire events. The fire investigator of the future is going to need a working knowledge of the science of fire, laboratory resources to make special reconstruction studies and the ability to apply computer models. These techniques bring an enhancement to evidence, and a clarity to elucidating critical issues.

Artificial Intelligence

Artificial Intelligence (AI) is an advanced form of computer technology that involves the programming of a computer to feature characteristics of human intelligence. The purpose of the program is to creatively solve problems with understanding and reasoning through experience rather than through the steps of a solution designed by the programmer. Programs may include the understanding of language, interpreting visual scenes, diagnosing diseases, analyzing electronic circuits, solving difficult equations, reasoning problems, and rendering expert diagnoses of conditions or situations.

AI began during the 1940s as a definition developed by Alan Turing who was a mathematician and logician. Mr. Turing stated:

> A machine has artificial intelligence when there is no discernible difference between the conversation generated by the machine and that of an intelligent person.

Two major forms of Artificial Intelligence that in the future will influence courtroom presentation of a fire scene are Virtual Reality and Expert Systems.

Virtual Reality (VR) is a computer-generated environment that is produced by utilizing a person's senses. The system is designed in two parts. The first part contains a helmet that completely surrounds the head. The function of the helmet is to block out the "real world" and replace it with the computer's generated 3-D lifelike environment through sensory equipment. The second part detects body movement that is analyzed through a glove or body suit to choreograph the user's position in the Virtual Reality. First used in the 1960s by the United States Air Force in flight simulators, VR has gained popularity with NASA and their space programs, an entertainment medium and used to promote commercially products on visual communication systems (television, video, movies, etc.). In the judicial system, VR will be used to present evidence through reenactments of the events. The crime scene can be explored from any angle, view, or perspective to analyze what occurred. This form of demonstrative evidence will be so lifelike, the trier of fact will actually feel the sensation of being present during the events in question. Experts agree that VR will advance with computer science and become a component of life in the future.

Expert Systems is the development of artificial intelligence programs used to help the judicial system be more consistent and efficient. The system works by taking a particular subject, program the knowledge and skills found in that field, develop a set of rules and a base of knowledge, then apply this information to a situation.

To understand how the Expert System works, we must first examine the basic concept of the expert witness in today's courts. Expert witnesses are individuals who testify and give opinions based on knowledge and skills not normally acquired by the average person. The trier of fact must then determine the validity of the testimony as part of his deliberation as to guilt or innocence. In the Expert System, the expert's opinion will be programmed into the computer to examine the expert's analysis and opinion.

One of the main problems facing ES is the computer's inability to employ abductive reasoning, or in simple terms, the human thought process. It may become necessary for the trier of fact to have an understanding of why the opinion was made, an accomplishment the computer, as yet, cannot achieve.

Artificial Intelligence, at the present time, is a new concept to be completely excepted in the judicial community. As new technology develops in computer science, problems will be eliminated so that AI will become an effective tool as demonstrative evidence.

Computer reenactments in many jurisdictions are ground-breaking technologies finding their way into the courtroom drama. Some lawyers and judges are unwilling to use computer animation in the courtroom, claiming the finished product is nothing more than a cartoon, or the special effects from a Disney production used to prejudice the jury. Other judges, in fear of repercussions, will refuse to have the admissibility of a new technology tested in their courtroom. The impact of computer animation in the courtroom is following the same obstacles that confronted the introduction and advancement of still photographs, motion pictures, and video that are today generally accepted forms of demonstrative evidence.

The future holds a bright promise for the relationship of computers and courtroom presentation. It is possible that within ten years, the judicial system will relax its interpretation and find the right criteria for accepting computer technology, such as Animation/Simulation, Virtual Reality, Artificial Intelligence and Expert Systems as demonstrative evidence.

SECTION 3

COURTROOM PRESENTATION

THE JUDICIAL SYSTEM

WHAT TO EXPECT IN THE COURTROOM

HOW TO PREPARE FOR TESTIMONY

DECIDING WHAT TO WEAR

COURTROOM DEMEANOR

CROSS EXAMINATION

THE EXPERT WITNESS

INTRODUCTION

A fire investigator may be notified at any time to respond to a fire scene for the purpose of determining the fire's origin and cause. Throughout the investigation, the scene should be documented and evidence taken to confirm the determination made by the investigator. This process is necessary for two reasons. First, the information gathered will assist the investigator in writing a report, and second, the documented information and evidence taken into custody may be called upon in the future for trial in a civil or criminal court.

The best practice to develop on any fire scene is to believe that *each investigation has the potential of going to court.* Keeping this philosophy in mind, each fire scene should be properly documented so that if the fire investigator is called upon to testify, all documentation will be in its proper place. A fire investigator who uses short cuts could find his or her testimony jeopardizing the case.

Chapter 13

KNOWING THE LEGAL SYSTEM

TYPES OF JUDICIAL SYSTEMS

The fire investigator's tree of knowledge of American Jurisprudence begins with a study of the judicial system. In the United States, the judicial system was established by the framers of the United States Constitution. After winning their freedom from England, the people of the thirteen colonies wanted a system of order and liberty that would not dissolve the new nation. The colonists wanted an independent form for government for each state as well as a central government. As part of the checks and balances to protect them from a type of government they left in England, Article III–The Judicial Article of The United States Constitution–was written. Article III states:

SECTION 1–The judicial power of the United States, shall be vested in one supreme court, and in such inferior courts as the congress may from time to time ordain and establish. The judges, both of the supreme and inferior courts, shall hold their offices during good behavior, and shall, at stated times, receive for their services, a compensation, which shall not be diminished during their continuance in office.

Jurisdiction

SECTION 2–The judicial power shall extend to all cases, in law and equity, arising under the constitution, the laws of the United States and treaties made, or which shall be made, under their authority; to all cases affecting ambassadors, other public ministers and consults; to all cases of admiralty and maritime jurisdiction; to controversies to which the United States shall be a party; to controversies between two or more states; between a state and a citizen of another state; between citizens of different states; between citizens of the same state claiming lands under grants of different states and between a state, or the citizens thereof, and foreign states, citizens or subjects.

In all cases affecting ambassadors, other public ministers and consuls, and those in which a state shall be party, the supreme court shall have the original jurisdiction. In all the other cases before mentioned, the supreme court shall have appellate jurisdiction, both as to law and fact, with such exceptions, and under such regulations as congress shall make.

The trial of all crimes, except in cases of impeachment, shall be by jury; and such trial shall be held in state where the said crimes shall have been committed; but when not committed within any state, the trial shall be at a place or places as congress may by law have directed.

Article III created two judicial systems in the United States: the federal courts and the state courts. Each of the courts can hear cases that are outlined in the United States Constitution. However, before a case can be heard in either the federal or state court, the court of original jurisdiction must be established. The court of original jurisdiction insures a court's authority to hear and determine a particular case in question. Once a case has been heard in the court of original jurisdiction and the outcome decided, the case may then be appealed by either party. This appeal is heard in a different court, a court that has Appellate Jurisdiction. Found in both federal and state systems Appellate Jurisdiction occurs when one of the two parties involved in the case seeks a review of the lower court's decision.

To understand the distinction between the federal and state court systems an examination of each will be discussed.

The federal court system is designed to rule on arguments pertaining to the United States Constitution and is divided into the following four principle courts:

District Courts

District Courts are trial courts that are courts of original jurisdiction in each of the fifty states, District of Columbia, and the Commonwealth of Puerto Rico. Cases include citizens from different states, enforcing or answering a question on federal law, cases between a citizen and an alien and civil suits over a certain amount. Figure 13-1 lists the location of each of the thirteen districts.

THE THIRTEEN FEDERAL JUDICIAL CIRCUITS

District of Columbia Circuit
  Washington D.C.

Federal Circuit
  Washington D.C.

First Circuit	Second Circuit	Third Circuit	Fourth Circuit
Maine	Connecticut	Delware	Maryland
Massachusetts	New York	New Jersey	North Carolina
New Hampshire	Vermont	Pennsylvania	South Carolina
Rhode Island		Virgin Island	Virginia
Puerto Rico			West Virginia

Fifth Circuit	Sixth Circuit	Seventh Circuit	Eighth CIrcuit
Texas	Kentucky	Illinois	Arkansas
Louisiana	Michigan	Indiana	Iowa
Mississippi	Ohio	Wisconsin	Minnesota
	Tennessee		Missouri
			Nebraska
			North Dakota
			South Dakota

Ninth Circuit	Tenth Circuit	Eleventh Circuit
Alaska	Colorado	Alabama
Arizona	Kansas	Florida
California	New Mexico	Georgia
Hawaii	Oklahoma	
Idaho	Utah	
Montana	Wyoming	
Oregon		
Washington		
Guam		
Northern Marianas		

Figure 13-1. The location of the thirteen federal judicial circuits.

Courts of Appeal

Courts of Appeal are courts that have appellate jurisdiction. Each of the eleven circuits in the United States reviews decisions of the District Courts within their circuit.

Supreme Court

The framers of the constitution created the Supreme Court, in part, so that the federal government could enforce and establish laws superior to the states, solve disputes between states, between an individual, and a state and between individuals from different states. The Supreme Court reviews the decision of the lower federal courts and the highest state courts. The process begins when a party petitions the nine judges of the Supreme Court to hear their case in what is called the "writ of certiorari." In order for the case to be picked, the "writ of certiorari" must pertain to some part of the United States Constitution. The Supreme Court picks the cases they want to hear based on the public interest. The Supreme Court is the last resort for a case to be heard in the federal or state system.

Specialized Courts

There are other federal courts that consist of specialized courts that deal with a particular subject. For instance, the Court of Claims will hear cases involving suits against the government of the United States. Other courts are the Court of Customs and Patent Appeals and Customs Court.

Any crime or law not defined in the United States Constitution cannot be heard in a federal court, and is therefore, heard through the state court system. Based on the same structure as the federal system, the state system also has original jurisdiction and appellate jurisdiction.

Original jurisdiction is usually divided into courts that try misdemeanor cases and higher courts that try both misdemeanor and felony cases.

Depending on the state, courts of original jurisdiction will have names like county courts, district courts, or superior courts, just to name a few. To know exactly the name of the courts in your state, consult the state's attorney's office in your district.

Appellate jurisdiction begins when a party seeks a review of their case in original jurisdiction. The case may be appealed all the way up to the state supreme court. If the case involves constitutional issues, the final appeal could be a "writ of certiorari" to the United States Supreme Court.

The majority of the fire investigator's courtroom testimony will be heard in civil and criminal courts. The function of a court is to establish a neutral area where two parties argue their differences and present their points of view in either a trial by jury or a nonjury trial. In a jury trial, the jury weighs the evidence, determines the creditability of witnesses, the credibility of the facts in issue, and lastly, comes to a conclusion based on what they heard. The judge is a neutral arbiter who makes sure the trial stays within the limits of the law. The judge will make a decision on all questions relating to law, decide whether or not evidence in the case should be admitted, and whether or not there is sufficient evidence to permit the case to be decided by a jury. A nonjury trial must be approved or denied by the court through a written request to waiver a jury trial. If granted, the judge performs his regular duties as well as those of the jury.

In civil courts, disputes are settled between persons, partnerships, corporations, or branches of government. In criminal court, a person is tried on charges of committing a particular crime that has been established by state law. If the individual pleads not guilty, then a defense is developed to establish his or her innocence. If an individual pleads guilty or is found guilty by the trier of fact, the criminal court will punish the guilty party for committing the crime.

Sequence of Events in the Courtroom

Subpoena

Before entering the courtroom, a witness will know of his or her participation by receiving a legal document called a Subpoena Ad Testificandum. The subpoena is issued in the witness' name by the court of original jurisdiction and it compels the witness to appear and testify before the court. A subpoena will vary in cosmetics depending on the jurisdiction issuing the writ. The subpoena should contain the following information: Jurisdiction issuing the subpoena, type of subpoena, name of the parties involved (In a criminal case either federal or state, the names will be United States or the name of the state, versus (or v for versus) the name of the defendant(s). In a civil case, the names will be the plaintiff who is the party initiating the case, and the defendant or defendants), date and time of appearance, location, and any other information the witness might need to know. Figure 13-2 is an example of a Federal Subpoena for an appearance in a Federal Civil Court. Exhibit A was a description of what the investigator was to bring into court. It stated:

EXHIBIT A

Any Fire Cause and Origin Reports prepared by you establishing the means and methods by which this fire was set, any videos or photographs taken of the fire scene (during and after the fire), any documents evidencing financial motive, such as business records, records of creditors and other financial records.

AO 88 (Rev. 11/91) Subpoena in a Civil Case

United States District Court

SOUTHERN _____ DISTRICT OF _____ FLORIDA _____

John Doe, Plaintiff

        V.

ABC Insurance Company, Defendant

     NAME OF WITNESS
TO: ADDRESS OF WITNESS

SUBPOENA IN A CIVIL CASE

CASE NUMBER: 94-0001 Div. Z
              Magistrate P.W. Brown
              June 8, 1994

[x] YOU ARE COMMANDED to appear in the United States District Court at the place, date, and time specified below to testify in the above case.

PLACE OF TESTIMONY	COURTROOM
East Courtroom– Second Floor Old Federal Building Anywhere, United States 00000	DATE AND TIME Trial Term Beginning June 20, 1994

[ ] YOU ARE COMMANDED to appear at the place, date, and time specified below to testify at the taking of a deposition in the above case.

PLACE OF DEPOSITION	DATE AND TIME

[X] YOU ARE COMMANDED to produce and permit inspection and copying of the following documents or objects at the place, date, and time specified below (list documents or objects):

SEE ATTACHED EXHIBIT "A"

PLACE	DATE AND TIME

[ ] YOU ARE COMMANDED to permit inspection of the following premises at the date and time specified below.

PREMISES	DATE AND TIME

Any organization not a party to this suit that is subpoenaed for the taking of a deposition shall designate one or more officers, directors, or managing agents, or other persons who consent to testify on its behalf, and may set forth, for each person designated, the matters on which the person will testify. Federal Rules of Civil Procedure, 30(b)(6).

ISSUING OFFICER SIGNATURE AND TITLE (INDICATE IF ATTORNEY FOR PLAINTIFF OR DEFENDANT)	DATE
Signature of attorney requesting the witness	

ISSUING OFFICER'S NAME, ADDRESS AND PHONE NUMBER

(See Rule 45, Federal Rules of Civil Procedure, Parts C & D on Reverse)

Figure 13-2. Face sheet for a federal civil subpoena

If the witness under subpoena does not appear at the appointed time, he or she may be found in contempt of the court and punished. In order to be excused, the witness must obtain some form of permission for their absence.

Another type of subpoena, called a Subpoena Duces Tecum (Figure 13-3), contains the same information as a Subpoena Ad Testificandum. The difference is that the Subpoena Duces Tecum is issued by the court on the behalf of one of the parties in the case and requires the witness to bring to court or to the hearing all documents or papers in their possession that are pertinent to the issues in the case. [NOTE: The fire investigator may not only receive a subpoena to testify at a trial, but also for grand jury hearings, motion to suppress hearings or any other pretrial hearing. The testimony in these hearings is recorded and transcribed, and can be used against the witness if the testimony differs. Make it a good practice to ask the attorney for a copy of your previous testimony to review your statements, reports, and production of evidence. This review should be conducted before entering the courtroom.]

IN THE CIRCUIT COURT OF THE FIFTEENTH JUDICIAL CIRCUIT
IN AND FOR PALM BEACH COUNTY, FLORIDA (CRIMINAL dIVISION)

CASE NO. _____

TO: name of witness
address of witness

STATE OF FLORIDA VS.

name of defendant

DUCES TECTUM: PLEASE BRING WITH YOU ALL DOCUMENTS, STATEMENTS, REPORTS, LAB RESULTS, PHOTOGRAPHS AND TAPES USED IN OR ACQUIRED DURING THE INVESTIGATION OF THIS CASE.

SUBPOENA FOR DUCES TECUM

GREETINGS:

You are Hereby Commanded to be and appear before a Judge of this Court at the Palm Beach

County Courthouse, Room _____, in West Palm Beach, Florida, on _____, 199__

at _____ M, to testify in behalf of the DEFENDANT wherein the State of Florida is prosecuting

and name of defendant is Defendant. Failure to appear will subject you to Contempt of Court.

Witness my hand and the seal of this Court on

_____date_____, 199__

NAME OF THE CURRENT APPOINTED CLERK

(CIRCUIT COURT SEAL)

BY:_____

name and address of attorney
issuing subpoena

Received this subpoena on the _____ day of _____, 199__, and excuted the same on the

on the _____ day of _____, 199__, by delivering a True Copy thereof to the within named

witnesses in the County of Palm Beach, State of Florida.

NAME OF THE CURRENT APPOINTED SHERIFF

By:_____
name of deputy serving the subpoena

Figure 13-3. A subpoena duces tecum.

Prior to and during the trial either party may file a motion with the court. In essence, a motion is a request made to the court for the purpose of obtaining a rule or an order to make a decision in favor of the party making the motion. The outcome of the motion is decided by the discretion of the judge. There are many types of motions in the judicial system. Two motions that affect the fire investigator the most are the motion to suppress evidence and the motion to suppress a confession. These two types of motions personally affect how the fire investigator conducted his or her investigation and determine if the evidence or confession was legally obtained. If the court finds the motion does involve illegal evidence or confession, that item is suppressed from being used in the trial. If evidence or certain statements are suppressed through a motion, the witness cannot bring up those subjects, or any subject or other evidence connected to the motio during the testimony of the trial. To do so can cause a mistrial. The granting of a motion to suppress may change the validity of the case.

In many states, there is a motion called discovery. This motion is a request to disclose all information that was gathered by the prosecution (or plaintiff).

This information can be facts, documents, written or recorded statements of witnesses, confessions of a defendant or codefendant, reports, tests, or any other form of information that has become a part of the case and is used to help prepare for the defendant's defense.

Procedures of a Trial

A typical trial by jury begins with the two opposing attorneys picking a jury through a process called voir dire. Citizens on jury duty are questioned by both attorneys on numerous subjects to determine if each individual jurist can serve on the jury with an unbiased attitude. The ultimate goal of voir dire is to have an impartial jury listen to the issues in the case and become fact finders before making a conclusion.

Once the jury has been selected, they are sworn in by the court and seated. The first thing the jury, or the judge in a nonjury trial, will hear is called opening statements. This is a speech given by each attorney to support the guilt or the innocence of the defendant. In a criminal case, the prosecutor will discuss the burden of proof through a brief detail of the evidence that will be presented, how this evidence will show a crime was committed, and how the evidence will prove the defendant guilty. In a civil case, the burden of proof must be established by the attorney for the plaintiff (the party initiating the suit) that a breech in a contract has been made or some type of wrong has been committed. The wrong, called a tort, can be either a civil wrong or an injury independent of contract, resulting from a breech of a legal duty.

The defense may or may not give an opening statement. If the defense decides to give an opening statement, it will be similar to the prosecutor's or plaintiff's with the exception of showing the innocence of their client.

At the completion of opening statements, witnesses are called to present the case for each party. In some instances, an attorney for one of the parties may ask the judge to invoke the rule, also known as Rule of Sequestration. If granted, witnesses cannot hear another witness testify and must wait their turn outside the courtroom. Second, witnesses cannot discuss their testimony with other witnesses until after closing statements.

In criminal cases, the prosecution must prove beyond a reasonable doubt the guilt of the defendant. The judicial system within the United States works on the principle that the defendant is innocent until proven guilty. This means the prosecution bears the burden of proof. For example, in a criminal arson trial, the prosecution must establish the presentation of evidence through witnesses that:

1. There is a corpus delicti–the body of the crime. In an arson case the corpus delicti is that a fire occurred.
2. That the fire involved the crime of arson (intent, willfulness and maliciously. The exact wording will depend on each individual state statute for arson.)
3. That the defendant is identified as the arsonist or had conspired and/or solicited for the arson to occur.

To prove their case, the prosecution (or the plaintiff in a civil case) is the first party to call witnesses. Through what is called direct examination, the attorney will ask questions of the witnesses, and through testimony, the witnesses will answer the questions. If the witness is an eye witness, the first questions will be about the witness. The second set of questions will deal with the chronological order of what the witness observed or did.

If the witness is an expert witness, the questions will first revolve around the qualifications of the witness (for further information on the subject, refer to Expert Witness, found later in this section). If the witness qualifies through the court to be an expert witness, the next series of questions will be based on the chronological order of the expert's examination. The witness will then be asked questions as to the opinion of the expert's conclusion.

Types of Evidence

In a trial, evidence is presented through the testimony of both the prosecution and defense witnesses. Evidence is best described as any medium used to prove or disprove a matter of alleged fact in a judicial trial. The three major classes of evidence used at a trial are direct, circumstantial, and real.

Through the verbal testimony of a witness, direct evidence is introduced through the personal knowledge of the witness to the facts in question using the five senses (hearing, sight, touch, smell, and taste). An example of direct evidence given by a witness: "I live across the street from the fire scene. Looking out my front window, I saw "Mr. Z" come out of his house carrying a red container by the handle. "Mr. Z" then proceeded to his neighbor's pickup truck and poured the contents of the container into the interior of the truck. "Mr. Z" then took a red stick in his hand and brushed the top of the stick onto the pavement. This started a bright flash that looked like a road flare. "Mr. Z" then stepped back from the open truck door and threw the flare into the cab of the truck. Immediately the vehicle exploded in flames."

Circumstantial evidence is not a direct fact, but rather a secondary fact that raises a strong inference to the facts at issue. An example of circumstantial evidence again deals with "Mr. Z". When he was arrested an hour after the fire incident was reported, his pants and shirt were seized through consent. Examined and tested by a forensic laboratory the clothes were found to contain gasoline residue. Having gasoline on his clothing raises an inference that "Mr. Z" may have committed the act. Circumstantial evidence, at times, may be called indirect evidence.

Real evidence can be any item or items that are associated with the offense. Real evidence helps place the defendant at the scene by identifying the item belonging to or in control of said defendant. Some examples of real evidence can be fingerprints, photographs, documents, human hair, type of weapon used, etc. In "Mr. Z's" case, real evidence was found and consisted of the container with "Mr. Z's" fingerprints on the container and a shoe print found in the soil by the open door of the vehicle. This print matched the sneakers worn by "Mr. Z" at the time of his arrest. The sneakers were also seized to be used as real evidence at the trial.

Lastly, before evidence is allowed to enter into testimony the foundation of admissibility must be met. This process involves preliminary questions asked of a witness to establish the witness' identification and qualifications along with the reliance of the evidence and authentication verified by the witness.

Walking into the Courtroom

The success or failure of an arson prosecution can depend on the quality of the witness' testimony and presentation of evidence. It will be at this time that the quality of work performed by the investigator will appear. Fire investigation reports, notes, photographs, videos, sketches, and physical evidence collected can be submitted as evidence along with the testimony of the fire investigator. Through the taking of complete and accurate notes, photographs, and sketches, this evidence will help bridge the gap between the investigation and the court case. Also, all of these forms of demonstrative evidence taken at the scene will become aids to help refresh the memory of the fire investigator. This is most important to a fire investigator who performs hundreds of fire investigations a year. Names, fire scene locations, dates and times, evidence, and classifications of the fire become a menagerie of information within the mind over a period of time. When testifying, the fire investigator will have to properly identify the evidence, explain where the evidence was taken, the date and time the evidence was taken, and who discovered and collected the evidence. The witness must be able to state that the evidence in court is the same evidence collected from the fire scene as physical evidence or is demonstrative evidence that depicts the fire scene as the fire investigator observed it. You can now see how notes, photographs, and sketches become a valuable asset to the fire investigator's testimony.

The bailiff will call the witness into the courtroom to testify. Remember, while walking through the courtroom doors, the jury's eyes are going to be watching the witness. His or her body language will develop first impressions and may have a direct cause and effect on whether or not they believe the testimony. Therefore, one should enter the courtroom as a strong figure with confidence, walking naturally with back straight, shoulders back, and head erect towards the witness stand. Approaching the stand, someone will direct the witness to look in the direction of the court clerk who will administer the witness' oath. One should turn and face the court clerk who will then ask the witness to raise his or her right hand. This is done with the hand being raised until the elbow is even with the shoulder. Fingers are closed together, pointing upward with the palm facing out. The clerk will then give the oath that will sound something like, "Do you swear that the testimony you are about to give today is the truth?" The witness will then answer in a loud and confident voice (not a scream or yell) "I do." Once the oath has been given, the witness can then be seated on the witness stand. The prosecutor will wait a few minutes for the witness to acclimate to the new surroundings. One should find a comfortable position in the chair and adjust the microphone towards the mouth so that he or she may be clearly heard by the jury. When one feels ready to give testimony, he or she should look up at the prosecutor. This will be the prosecutor's sign the witness is ready for direct examination.

When the prosecution (or plaintiff) has completed direct examination of its witnesses, the attorney will say to the judge, "I have no further questions, your Honor." The judge will then acknowledge the defense attorney and ask if they have any questions for the witness. If so, the defense attorney may now ask the witness questions to try to discredit or lessen the damage caused by the witness. This type of questioning is called cross-examination. This type of examination is an option to the defense. When the defense concludes his or her initial questioning, the judge will be informed by the phrase, "I have no further questions, your Honor." The prosecutor will be asked by the judge if there are any other questions by the prosecution (or plaintiff) on what is called redirect examination. If the prosecutor redirects, then the defense is also allowed by the court to conduct a recross-examination. (Redirect and recross are options to be used by the attorneys.) When both parties have exhausted their questioning of the witness, the judge will dismiss the witness. At no time during or after giving testimony can the witness discuss with a jurist or another witness the facts in the case. This is called invoking the rule. Another fact to consider as a witness involves a court recess in the middle of testimony. Again, any discussion with a witness, jurist, or even any of the attorneys concerning the facts of the case must be avoided. Any violation of these rules can cause a mistrial. Discussion of the case should be done after the decision of the case has been announced.

The prosecutor (or plaintiff) may ask the court, before the witness is excused, to keep the witness on stand-by as a rebuttal witness. A rebuttal witness is used to refute or disprove the evidence and/or testimony of the adverse party. Once the prosecution (or the plaintiff in a civil trial) has completed their presentation of witnesses and evidence, the prosecution or plaintiff will state that they rest their case.

The focus of the trial now turns to the defense. The attorney, before calling any witnesses to the stand, may ask the court for a motion of an immediate dismissal, based on the fact that the prosecution (or plaintiff) failed to prove their case. If the motion is granted, the trial is completed and in favor of the defense. When the motion is denied, the trial continues and the defense will begin to present their side of the case.

The defense will call witnesses to the stand, like the prosecutor (or plaintiff), to present evidence that will show their client as being innocent of committing a crime, or in a civil case, of committing a "wrong." The same procedures apply to the defense witness as it did to the prosecution (or plaintiff) witness. The witness is sworn, seated, goes through direct examination with the defense attorney, may possibly go through cross-examination with the attorney for the prosecution, and may go through redirect or recross examination. After the completion of all testimony of the defense witnesses and the presentation of defense evidence, the defense will rest.

After the completion of the defense side of the case, the judge will then ask the prosecutor (or plaintiff) if there is any rebuttal. If so, the witness is called and goes through the process of examination again. The only difference is that if the witness has already been sworn in, then he or she is reminded that they are still under oath. At the completion of rebuttal, the defense may counter and produce a witness for their own rebuttal, called surrebuttal.

After the completion of all the testimony from the witnesses and the presentation of evidence, each side will have the opportunity to talk to the jury. This is called the closing arguments or sometimes known as the final argument or summation. Each attorney will also try to discredit the other side's witnesses and evidence as a ploy to bolster their justification.

After closing statements, the judge will read to the jury instructions that may contain some of the following: the applicable law concerning the case, the burden of proof, responsibility of the court and jury, and a reminder that the defendant is innocent until proven guilty. It is up to the discretion of the judge as to what verbiage will be used in the instructions.

After hearing the instructions, the jury will go into seclusion in the jury room to deliberate the defendant's outcome. In criminal court, if the defendant is being charged with more than one criminal count, then the jury must deliberate for each individual count.

If all the jurists cannot come to an agreement on their decision, then a hung jury is announced by the judge and the complete case may be retried. When the jury has made its decision, they are brought back into the courtroom where their decision, called a verdict is read. In a nonjury trial, this decision will be made by the court.

Chapter 14

TESTIFYING AS A WITNESS

To be a successful witness, the investigator must not only have the experience, technology, and terminology of fire scene investigation, but also must possess the ability to present evidence in a convincing manner through testimony as an expert witness. It should always be in the back of the witness's mind that each juror must decide if the witness is believable or being deceiving. To help resolve the juror's dilemma, a positive or negative impression of the witness may be formulated. The juror will look at the witness's appearance, body language, and the way the testimony is presented. Therefore, it is in the best interest of the fire investigator to be prepared in mind and body. The unprepared investigator can be the most destructive force the individual faces.

Preparation of the mind begins immediately after receiving the subpoena. A review of the file should be made so that the investigator gets reacquainted with the particular details of the case. After reviewing the case, a telephone call and follow-up visit with the attorney should be made.

Always try to arrange a meeting with the attorney involved on your behalf. In a criminal case, the visit will be with the prosecutor known as the assistant state's attorney, assistant district attorney, or the assistant prosecuting attorney. This meeting should be conducted before the witness appears in court. If the prosecutor has not made contact with the investigator, then it is up to the fire investigator to initiate the meeting. The meeting establishes a communication between the fire investigator and the prosecutor and gives the prosecutor a personal knowledge of the fire investigator. Further contact will enable the prosecutor to know the qualifications and limitations of the investigator.

Since the prosecutor will be asking some very important questions during the trial, it is a good idea to go over the case together to find and acknowledge the strengths and weaknesses of the case. At the meeting, the prosecutor can examine the evidence. Photographs and sketches can be discussed in detail to determine which items are going to be used as demonstrative evidence. In some instances, this will be the first arson case the prosecutor will handle. Therefore, it is the responsibility of the fire investigator to educate the prosecutor in the procedures and definitions used during a fire investigation.

In a civil case, the fire investigator may be testifying for the plaintiff or the defense. In either case, a meeting should also be arranged. Similar to the meeting with the prosecutor, a meeting with the civil attorney enables both parties to go over the investigator's role in the investigation and to determine what questions might be asked.

The night before testifying spend some quality time re-reading all written reports made during the investigation. If prior testimony was given concerning the same case (a deposition or a pretrial hearing), then a review of those transcriptions should be made to ensure the same testimony will be given at the trial.

Reviewing photographs and/or videos will prepare the investigator in several areas. First, a review will fortify his or her memory as to how the overall fire scene appeared and orient the investigator to the different areas of the scene. Second, if the photographs were properly taken, they will assist the investigator in developing the progression of the fire scene investigation. Third, photographs can benefit the witness in recalling the fire scene indicators used to make the fire's determination. Lastly, a review of photographs depicting the location of evidence will be fresh in one's mind for the day of testimony.

Another solid reason to review the facts in the case is to avoid using notes or reports as a continual reference while testifying. To do so gives the opposing attorney the right to examine the material, and if warranted, ask questions. Also, if the witness is not prepared and tries to give testimony through the reading of his or her report, the defense may object to the reading. This does not mean that the case file must be left at home. Instead, all documents should be placed neatly in a folder in known chronological order. In this way, if a question is asked and the immediate answer is not known from memory, the witness may ask the court for permission to refer to his or her file. Testimony from notes or reports cannot be given unless permission is granted by the judge or counsel.

The last item to go over is the witness's most up-to-date version of his or her resume. This document should be an ongoing written history of the investigator's career history. The information should include employment history, education, and special schooling and organizations belonged to.

Deciding What to Wear

As silly as it may sound, what a witness wears in the courtroom can influence a jury in their decision. R. Rogge Dunn said it best in his article *"Persuasive Expert Testimony:*

> Appearance definitely shapes the credibility of what one says. If you make a good first impression, the jury will like you or respect you and give more credence to your testimony.

The overall object of the witness is not to bring the jury's attention to the witness through appearance, but rather to have the focus of attention concentrated on what the witness is conveying in testimony. A neat and conservative appearance will prevent the jury from possibly typecasting the witness. Objectionable styles of clothing, especially in woman's wear, may offend the sensibilities of some jurists. Clothing with bright colors that may be appropriate for an evening out would likely be most inappropriate for the courtroom. Clothing that looks obviously expensive may offend jurists with low income. Flashy jewelry has a tendency to distract the jury and should not be worn. Any jewelry worn at all should be limited and conservative.

The type of clothing worn is only half the battle of appearance. The other half centers on the witness' grooming standards. For men, hair should be neatly combed, hands and nails clean, and face shaved or beard/mustache trimmed. Clothes should be neatly pressed and shoes shined. Do not display fraternal pins; lodge or club emblems should not be displayed on clothing. Shirt or jacket pockets should not bulge out. Items such as cigarette packs, small note books, or any other material that could cause a distraction should not be brought to the stand. Keys and coins should be placed securely so that they do not jingle. A nervous witness may have a tendency to play with items such as the keys or coins in their pockets, again causing a distraction.

If one wears an official uniform during normal working hours, then it should also be worn for the courtroom appearance. Members of the jury feel comfortable around individuals in uniform since it stands out as a sign of authority. To add sparkle to the uniform, it should be cleaned and neatly pressed. If a coat is not part of the uniform, then one should be especially careful of what is in the shirt pockets.

A male witness who is not in uniform should wear a suit or dress coat and tie. The color of the suit or jacket should be a subdued color, preferably a dark blue. The shirt should be white with a conservative tie. Fancy ties should be left for social gatherings.

The female witness who is not in uniform should also wear a dark color suit or a conservative dress. The length of the dress should be approximately knee length or longer. Short skirts and low cut blouses are improper attire for a professional witness. Jewelry should consist of simple earrings and a simple nonflashy necklace. Flashy rings should be avoided except for engagement and wedding rings.

After reviewing the case and selecting clothes for the courtroom appearance one should get a good night's rest. The witness should awaken refreshed and prepared to testify as a fire investigator witness.

Day of Court Appearance

As the time approaches for the scheduled appointment in the courtroom, "butterflies" may begin to set in. Signs of nervousness will begin to appear. A body can begin to produce adrenaline that will increase the heart rate, making breathing shallow and fast (causing a dry mouth) and the development of different stages of perspiration (sweaty hands, beads of sweat on the forehead, etc.). As the time approaches closer for the appearance on the witness stand one's mind will begin to play deceitful games. In going over the case for the final time a witness may forget some significant elements of the investigation or forget basic terminology and procedures of a fire investigation that at any other given time would be known forward and backwards. One should relax! These are all normal reactions that investigators face in one form or another.

There are many methods that can be used to overcome these obstacles. First, it is wise to avoid caffeine products as they tend to increase the already heightened apprehension of testifying.

Second, one should arrive at the courthouse early. The extra time can be used to go over all of the notes and reports. There are very few accepted excuses for being late. By being rushed, the investigator witness will appear disorganized. The precipitate mood of the investigator will even carry over to the way he or she responds during testimony. Therefore, tardiness is uncalled for, very unprofessional, and will only help to defeat the investigator.

Third, individuals have a tendency to overstudy a subject, causing mental fatigue. Before one's name is called to enter the courtroom it is good to take a few moments to relax instead of cramming in last minute details that by now should be cognizant. The elementary concept to relaxation is through breathing. A breathing exercise of slow deep breaths and expanding the abdomen region can be helpful. Breath should be held for a few seconds, then slowly exhaled, contracting the abdomen region, and repeated several times. Throughout this exercise one should think of something cleansing, pleasing, beautiful or joyful. For example, while taking in deep breaths it is helpful to focus on erasing everything from the mind by saying to oneself "clean mind," and when exhaling, saying "alert body." To some a pleasing thought would be to envision sitting by a mountain stream and hearing the cool water cascade over the rocks. Selecting a thought works best, implementing the breathing exercise, purgeing all other thoughts will bring about a sense of relaxations.

When the moment arrives and the witness' name is called by the bailiff, all eyes will be focused upon him or her. The testimony will become the focal point of attention for the trier of fact. One should not become overconfident approaching the witness stand. As the fire investigator witness, it is important to understand three characteristics about the trier of fact. One, in all probability, the trier of fact knows nothing about fire investigation. Two, the trier of fact does not comprehend the terminology used by fire investigators to describe a fire scene, and three, the trier of facts will have very little knowledge in the field of fire science. One's testimony as a fire investigator witness will educate the trier of fact in these areas. The presentation produced on the stand should communicate to the trier of fact that the witness, as a fire investigator, is intelligent, knowledgeable, truthful, unbiased, and that his or her testimony is a clear fact-finding statement. One should never assume that the trier of fact knows what he or she knows. In all probability the trier of fact knows very little about the fire investigation and knows even less on how the investigator came to the determination of the fire. The trier of fact achieves this knowledge *through* the testimony of the fire investigator.

To get the facts across to the trier of fact each attorney will ask the witness questions. It is essential to listen carefully and give the attorney full attention as the question is being asked. The witness should not give an immediate answer after the completion of the question but rather take a few seconds to think over the response. This slight delay also gives the opposing attorney time to maneuver with an objection. If an objection is heard the question should not be answered. The judge will hear the objection then make a decision. If the judge overrules the objection, the witness will be instructed to answer the last question. If the judge agrees with the objection, he or she will say "sustain" and the attorney must rephrase the question before the witness can give an answer. However, one should be cautious on the duration of the pause. A long hesitation might give the jury the impression of a deceitful answer.

The fire investigator witness should always tell the truth. A truthful investigator will be able to give his or her testimony over and over, in any chronological order and in so doing will withstand the opposing attorney's tactics to twist or change the truth. It is important not to fabricate or distort stories by giving half-truths. If the answer is not known then the response should be, "I don't know" or "I can't recall." One lie will develop into more lies until the lies can no longer be remembered. Eventually, the witness will be caught in the lies and find his or her creditability tarnished. It can take only one lie to ruin the reputation of a fire investigator and end a hard-working career.

More helpful hints before answering questions:
- Do not anticipate an attorney's question and start to give an answer before the question is completed. This is impolite, and second, this quickness could lead you into a trap.
- If you do not understand the question, ask the attorney to repeat the question. There is nothing wrong in stating "I do not understand; could you please repeat the question?" CAUTION: Used often, this phrase could be taken as a stalling tactic.
- If you are confused in the way the question is presented, ask for it to be repeated in a different way.
- If the attorney refuses the witness' request to repeat the question, you may turn and with respect ask the judge if the question can be repeated.
- If the witness discovers that he or she gave an incorrect answer while on the stand, the witnes should clarify the mistake as soon

as possible. It is better to correct the wrong instead of having the opposing attorney correct it.

- Rehearsing testimony will not sound genuine. This type of testimony gives the trier of fact a false or fake impression of the witness.
- Listen to the question and give only the answer to that question. Do not volunteer or give further information.
- At times, attorneys will request an answer in a "yes" or "no" form. There will be times when the witness will feel a need to qualify the "yes" or "no" answer. Before answering, the witness should state the answer needs an explanation. If the attorney is on your side, he or she should be alert enough to ask you to qualify that answer.
- When showing and testifying to an exhibit of the court, verbally refer to the exhibit number first, for identification, then proceed with testimony. An example, "court exhibit #5, represents a photograph depicting....

Helpful Body Language Tips

To get ahead in the testifying game, there are many conscious and subconscious behaviors that can enrich the performance of a witness on the stand. The first group of behavior traits concentrates on the witness' appearance and how the witness conveys his or her composure while walking up to the stand (discussed earlier in this section). The second group of behavior characteristics centers on the courtroom demeanor of the witness as he or she is giving testimony.

In front of an audience, a speaker effectively channels his or her message through three methods: voice, type of words spoken, and body language. How these methods are projected determines the response from the audience. A poor speaker can put to sleep a good part of the audience with a monotone voice, and lose another part of the audience by using terminology foreign to the listener. The remainder of the audience's attention span wanes due to body language that resembles a piece of cardboard, revealing very little character of the speaker. An excellent speaker will hold the attention of the audience with a dynamic voice, speak to the audience in words they understand, and through body language, make each member of the audience feel like he or she is being talked to individually. The fire investigator witness can produce the same techniques of a good speaker by developing and practicing communication skills and projection of body language.

Familiarizing voice projection is probably the easiest to learn of the three methods of courtroom demeanor. The witness should speak clearly and loud enough to be heard throughout the courtroom, being forceful but not to the point of straining. A low voice can be interpreted as a nervous trait or a sign of deception. One should not get in the routine of using the same pitch or rhythm. A speaker, using the same pitch and cadence, will quickly lose the attention of the listener. A witness should raise his or her voice to emphasize important issues, then pause to let the words settle with the listener, and take a short pause between sentences. Sentences should not be connected with with and, ah, umm, etc. Listening to a radio or television announcer can be helpful.

When answering an affirmative or a negative question with one word, one should complete the answer in a professional manner (such as yes, yes ma'am, yes sir, no, no ma'am, or no sir). Slang responses of umm hmm, yeah, yep, ah ha, nah, and nope, cheapens the character of the witness. A witness should always be respectful when addressing either attorney by referring to them by their last name or as sir or ma'am, and never by their first name. This indicates a personal relationship that should be left out of the courtroom. Above all, the judge is address as "your honor," a sign of high esteem.

Many individuals are in the habit of stalling for time when asked a question by using nonwords. Phrases such as "ah," "aha," "uh," or "I, I, I mean," or any other partial word, are clues of nervousness or deceit. There are many ways to overcome these nonwords. First, a person should listen to his or her voice on a tape recorder. Second, another person can and every time a nonword is spoken, ring a bell. Third is listening to oneself speak and make a conscious effort to eliminate nonwords. Fourth, by taking a speech class, the curriculum will develop methods to overcome these and other negative speech patterns.

The second method, type of words spoken, involves how the witness talks to the jury and or judge. Most people pay attention to the spoken word because it is the easiest way to communicate between individuals. Throughout our lives we have been taught to listen closely to what is being said. Based on this experience, the spoken word becomes more accountable than that of the voice or the way body movement is translated. The speaker can transmit messages easier through words than by face or body language. Many expert witnesses, as a form of communicating their knowledge or skill, will use technical terminology as a way of impressing the jury. Without an explanation, the lay person in the jury box will only find the testimony confusing and lose the important concepts expressed by the witness. The best approach is to keep in mind that the jury consists of men and women who have a variety of backgrounds and education. To reach each individual jurist, it's best to speak in simple everyday language. If fire terminology is used, the term should be illustrated in plain descriptive language. One should never talk down to the jury. Jurors have a tendency to dislike "know it alls" or individuals with cocky attitudes. Remember, the jury is there to hear the facts in the case. If the jury cannot understand the facts, then they cannot make a decision.

The third method is a form of nonverbal communication called body language that is used in combination with voice and the spoken word. The concept centers around the idea that each individual, when spoken to, not only listens to what the speaker is saying but also watches and reacts to how the body of the speaker moves when talking. The speaker will transmit these nonverbal signs voluntarily and/or involuntarily. (Involuntary body movement is caused by the autonomic nervous system. In almost all cases, the movement cannot be consciously regulated. Blushing, getting red with anger, sweating, breathing, and the heart rate are some examples of the autonomic nervous system at work.) The listener watching the speaker will interpret the projected signs consciously and/or subconsciously. A good example of nonverbal communication can be observed at a clown's performance. Through facial expressions and other exaggerated forms of body movement, the clown can manipulate the audience to be happy or sad, without saying a word.

Compared to the other two methods, voice and the spoken word, body language is the hardest to learn and can take a lifetime to master. This learning difficulty is not as hard as it sounds. The fire investigator witness who is a novice in body language behavior can begin learning to project and interpret some basic skills that will be useful on the witness stand. With time and practice these techniques can be applied by the investigator as a listener during interviews and interrogations. (In this role, the investigator is searching for signs of deception during questioning.)

One of the easiest signs of nonverbal communication to observe is found in the face. This is especially true of the courtroom witness, who is seated behind a partial partition on the witness stand, with the only visible part of the body being from the waist up. Through facial expressions, different emotions such as, but not limited to, happiness, fear, anger, sadness, excitement, surprise, contempt, and disgust can be displayed in three primary areas of the face: the eyebrows, the eyes, and the mouth.

The raising and lowering of the eyebrows are used as conversation signals in conjunction with speech. Raised eyebrows can indicate a question, disbelief, skepticism or an exclamation; while lowered brows or bringing the brows together reveals perplexity and concentration.

The second area, the eyes are known as "windows of a person's soul." The eyes can reveal the inner-most true feelings of an individual. The direction the speaker is looking can tell many things about his or her mood. If, when questioned, the speaker is looking down, the feeling could be sadness. When looking up or to the upper left or right, the speaker is thinking. Looking down and away is the expression of feeling shame or disgust.

To help get the point across to the jury, the witness should develop the habit of periodically making a visual contact with each individual juror. This eye contact creates a personal bond between, the witness and the jury. At different times the witness should look at a different juror, makeing the juror feel that he or she is being talked to directly. Avoiding the jury's eyes could subconsciously make them feel the witness is hiding something. The witness must also be careful on the duration of the gaze. Holding the gaze too long could create the feeling of being dominated and will make the listener uneasy. The best approach to use centers on the attorney's questions. If the answer to the question is a short reply, one should gaze at the attorney. When the answer is longer, that is the cue to look at the jury and then answer the question. Many opposing attorneys will catch this maneuver and will walk from the jury box to the opposite side hoping to draw the witness's attention from the jury.

Another reason to keep a visual on the jury is to look for signs of encouragement. As the witness gives the testimony, members of the jury will subconsciously nod their head or make soft sounds like "umm-hmms" as a positive reaction. This indicates the witness's point is being recognized and understood. Negative signs such as daydreaming, sleeping, or looking somewhere else are indications that the witness needs to change his or her presentation. When confronting negative signs, all is not lost. One should regain composure and analyze his or her testimony up to the point of discovery. The witness must think simple. Unless there is a recess, or some form of a break in the testimony, the witness cannot call for a time-out. Questions are still being asked that need to be answered. It is important to remember the three methods for courtroom demeanor. Has my voice been too monotone? If so, raise it to stress an important point. Has my fire terminology been too technical? If so, bring it down to layman's language. The witness should talk to the jury as if it were everyday conversation. Am I so nervous that I might appear to look like a cardboard figure? If so, the use of body language can be helpful. One should learn to relax and express emotions through the body. Once the flaw is detected and corrected the jury should be reevaluated. The results could be amazing!

The third area of facial expression is the mouth. Three major emotions are exhibited by the mouth. A frown that turns the ends of the mouth downward indicates sadness. Anger is revealed by the narrowing of the lips and may be done in conjunction with the narrowing of the eyes. The smile turns the ends of the mouth upward and indicates a form of pleasure or contentment.

The smile is the most frequent of all facial expressions to be observed. It takes but one muscle in the face to show enjoyment, while most of the other emotions need three to five muscles. Almost all individuals respond favorably to someone that smiles. A smile will often be returned.

The fire investigator witness, during testimony, will be using and looking for a "coordination smile." In the book *Telling Lies*, Paul Ekman states that a coordination smile:

> Regulates the exchange between two or more people. It is a polite, cooperative smile that serves to smoothly show agreement, understanding, intention to perform, acknowledgment of another's proper performance. It involves a slight smile, usually asymmetrical, without the action of the muscles orbiting the eyes.

Another type of coordination smile is the listener smile. Paul Ekman states that the listener smile is:

A particular coordination smile used when listening to let the person speaking know that everything is understood and that there is no need to repeat or rephrase.

This type of smile can also be used in conjunction with the head nod as a positive indicator that the witness's testimony is being understood.

It is important to understand that courtroom testimony is a solemn and serious matter. Oversmiling or laughter, taken out of context, could indicate deceit, a biased attitude, a "I don't care" attitude, or an overconfident, know-it-all, cocky individual, all of which represent the demeanor of a nonprofessional witness.

The best place to project body language is when the witness is asked to step down from the witness stand and explain an exhibit (evidence) of the court. In most cases, this part of the testimony will take place in front of the jury. Posture will become an important nonverbal communication skill to project. An erect body looking at the jury from a close distance will display confidence, give voice projection with little effort, and allow the witness to observe the reactions of the jury.

As stated earlier, there are times when jurors will lose their concentration. Utilize this time to regain the attention of those lost jurors. If a juror is found not attentive or dozing off, the witness should move ever so slightly to an area in front of that juror. This slow movement will seem natural and not cause embarrassment to the inattentive juror. Once in position the witness should raise his or her voice a bit and look at that individual. The wandering juror will hear the reflection in the voice and become conscious of the witness standing directly in front. The witness should briefly meet and fix on the juror's eyes. The juror now knows that he or she has been discreetly singled out and should give the witness his or her undivided attention.

The same technique applies to the dozing juror who becomes aware, subconsciously, of a presence out of the ordinary. Waking up from his or her light sleep, the juror will become cognizant of the witness's brief stare. The witness now has the attention of that juror.

Working the jury will keep their attention span high. The witness should make it appear that he or she is talking to each individual member of the jury, verbally and nonverbally through body language, to make each one feel like he or she is a part of the case.

During the witness's time in front of the jury, there are certain things that should not be done. The following activities only create distractions and/or give the witness the appearance of being very nervous.

- Fidgeting with eyeglasses
- Pulling on an ear lobe
- Wearing sunglasses on face or head
- Playing with jewelry
- Playing with objects in hands
- Playing with buttons
- Putting hands in pockets
- Wringing hands
- Tapping with fingers or feet
- Chewing gum

When showing and testifying exhibits of the court, one should always explain in detail what it is being referred to. The purpose of the detailed description is to have it recorded by the court reporter. When the testimony is transcribed, the exhibit will be clearly explained through words. For example, if the witness was pointing to a photograph and stated, "This is where the fire started," the only record of the testimony would be, "This is where the fire started." The proper way would be, "court exhibit number 5 is a photograph depicting the southwest corner of the master bedroom. In the lower right hand corner of the photograph is a duplex electric outlet with heavy charring to the wood paneling surrounding the outlet. Extending from the tip of the outlet are two copper electrical wires with heavy beading. Directly below the outlet on the floor are the remains of the window curtain. The fire's point of origin was determined to be the electrical duplex where an electrical short ignited the window curtain."

More Information on the Cross-Examination

Cross examination is a series of questions asked by the opposing attorney that pertain to the creditability and testimony given by the witness during direct examination. (On rare occasions the courts have allowed opposing attorneys to deviate from the direct examination testimony if they can give a valid and just cause for their action. The judge will rule on the matter and, if in agreement, will allow the attorney to continue under strict scrutiny.) In general then, if the witness has properly conducted the investigation, has taken the time to prepare for the trial, and has told the truth during the direct examination, the attack on the witness during cross-examination will be minimized.

The main purpose of a cross-examination is to either lessen the damage caused by the witness during direct examination or to attack the creditability of the witness. The opposing attorney will usually cover the most devastating material first for two reasons: (1) The witness will be more nervous and uneasy at the beginning of the cross-examination and (2) The jury's attention is greater at the beginning and with time wanders.

To help prepare the witness for cross-examination, he or she should know some of the legal tactics and techniques used during cross-examination. There are four areas in which the opposing attorney will attack.

1. The cross-examination will make the witness appear that he or she is lying. Throughout this section, an important theme stressed has been honesty. The witness gets into trouble by telling outright lies, stretching the truth by telling half truths or making assumptions.

2. Mistakes were made during the investigation. Fire investigators are human and therefore will make mistakes. If a mistake has been brought to the attention of the witness and the witness can honestly see the mistake, then he or she should admit to it. Admitting to small mistakes will do very little damage to the witness's creditability. Most jurors will sympathize with the witness. Trying to cover up mistakes opens the door for a prepared opposing attorney. To save face, the witness turns to lying. The coverup is not worth going to jail for perjury and the downfall of a career.

3. The attitude of the witness is biased. It is important for the witness to remain neutral throughout his or her testimony. In his book, *Taking the Stand*, Dennis Tilton states:

> A trial witness is supposed to report observations and actions that may help the jury to make informal and just decisions. A witness is not meant to be an advocate for guilt or innocence. To keep credibility intact, he or she must seek to protect an attitude of strict and absolute neutrality in court.

Mr. Tilton has developed some questions that the witness might hear in an attempt to dislodge the neutrality of a witness. Following the question is Mr. Tilton's proper answer.

Q: "You dislike my client, don't you?"
Inappropriate A: "Yes, I do."
Appropriate A: "No, sir."

Q: "You want to see my client convicted, don't you?"
Inappropriate A: "I certainly do!"
Appropriate A: "That's for the jury to decide. I have no deep-seated feeling about that one way or the other."

Q: "You'd like to see the defendant here punished, wouldn't you?"
Inappropriate A: "I definitely would. He deserves it."
Appropriate A: "That's up to the judge, sir Not to me. I've given no real thought to possible punishment."

4.   The investigation was not conducted properly. All investigations should be conducted completely and properly within the limits of the law. Short-cuts will only lead to trouble. An investigator will never know at the fire scene if that particular case is going to court or not. It will take at least several weeks after the scene investigation to make the determination. The best policy a fire investigator witness can demand of him/herself is to treat each and every fire scene as if it was going to court.

Other Do's and Don'ts of Cross-Examination

• When answering a question make it simple and to the point. Volunteering additional information gives the opposing attorney further information, more questions to ask the witness and more time to turn the jury against the witness.
• Be as polite and courteous as possible. The opposing attorney will try different tactics to make the witness combative, angry, and argumentative in an attempt to turn the jury against the witness.
• Never, during cross-examination should the witness look or use nonverbal gestures towards his or her attorney for encouragement or help.
• If interrupted by the opposing attorney before finishing, the witness can politely inform the judge that he or she would like to finish the answer.

- Attorneys like to ask trap questions to confuse the witness and make him or her think something is there behind the question. It is a tactic used to rattle the witness. A sample questions would be, "Did you ever discuss this case with anyone since receiving your subpoena to appear in court?" The answer, of course, is "Yes" and don't be afraid to name the individuals, including your attorney.
- If the witness's answers are quick on direct and slow in coming out on cross, the jury will pick up on the hesitation and feel the witness may be in trouble or about to be deceitful.
- If the witness does not know the answer, her or she should say so. One should not make up an answer or assume one.
- If one hears the question, "Do you want the jury to understand..." BE ON GUARD! Listen to the question or statement very carefully. If it is not exactly what you want, clear it up, then give the answer.

Remember, to properly prepare for court and feel the experience of a successful conclusion, the witness should:

1. Know what is expected of the witness in the courtroom.
2. Review the case with the attorney prior to the trial date.
3. Review all notes, written reports and one's career resume.
4. Review all demonstrative evidence to be presented.
5. Make a conscious effort of personal appearance.
6. Learn the proper etiquette for a witness while in the courtroom.
7. Only testify to the knowledge and actions performed by the investigator and maintain an attitude of neutrality.

Chapter 15

THE EXPERT WITNESS

The cornerstone of any fire investigation in litigation will be the testimony of the fire expert witness. As an expert witness, it can be the most critical role for the fire investigator to perform in the courtroom. Regardless of the type of trial, civil or criminal, the fire expert witness may fulfill a number of functions during the trial. In his book, Fire Litigation Handbook, Dennis J. Berry cites five different roles that the expert may be involved in. They are:

1. The cause and origin expert who examined the scene of the fire will introduce essential facts developed from fire scene photographs and physical evidence obtained from the scene, and offer a narrative description of the fire damage.

2. The cause and origin investigator/expert also plays an essential educational role in explaining to the jury the logic and methodology of the proper investigative techniques used in determining cause and origin, and ties together the circumstantial facts which lead to his or her conclusion as to the cause and origin, or spread of the fire.

3. The testimony of the cause and origin expert is also the essential predicate to the presentation of other scientific expert testimony.

4. The cause and origin expert may be required to present evidence that negates certain theories of cause and origin.

5. Experts in fire cases serve an educational function for counsel.

The fire investigator witness may be asked to act upon all five or a combination of the different roles. The witness may be asked to intro-

301

duce through testimony, and usually in a chronological order, the facts and evidence documenting the fire investigation. During this testimony, the fire expert witness will have to educate the trier of fact on how the systematic approach, terminology, and fire indicators were used to make a determination into the fire's cause. As each topic appears, stop! Take the time to explain in everyday language the terminology and/or methods used, then continue with the testimony. Once the particular facts and evidence have been admitted as testimony, the fire investigator expert witness, depending on the discretion of the judge, may give an opinion as to the origin and cause of the fire.

There will be times when the cause of the fire is beyond the fire expert's knowledge. In this case, the fire investigator expert witness will lay the groundwork for a scientific expert to verify the fire's cause. For example, the fire investigator testifies about the fire scene and how a systematic approach was used to determine the origin and cause. At the origin of the fire, near the point of origin, the fire investigator takes into evidence a damaged electrical appliance. The fire investigator's opinion is that the electrical appliance started the fire. However, because the fire investigator has limited experience in the electrical field, an electrical engineer will be used as an expert to determine the appliance's failure.

The same concept also applies to a fire that has been determined to be incendiary through the use of what the fire investigator believes to be a flammable/combustible liquid. Samples were taken into evidence and determined through laboratory analysis to be gasoline. The fire investigator will testify to the fire scene, methods used in the investigation and the indicators used to help determine the fire. This testimony lays the foundation for the scientific expert, the chemist, to testify on the flammable/combustible evidence.

The fire expert witness will at times find him or herself in opposition to the adversary's expert witness. Each expert will base their conclusion, in part, by listening to witness's statements, observing patterns of fire behavior, and through a process of elimination, determine the fire's cause, each with a different causation. The importance of presentation, appearance, and demeanor now becomes critical as to whose testimony is the most believable.

Fire investigation is a specialized field that many attorneys do not comprehend. Without the knowledge of the fire investigator expert, a criminal or fire litigation case will be defeated. To the rescue comes the fire investigator expert. First, the fire investigator expert will instruct the attorney in the proper techniques and vocabulary of a fire investigation. Second, the fire investigator expert will help find weaknesses in the adversary's theory of fire cause and will assist the attorney in forming cross-examination questions. Lastly, the fire investigator expert could be called upon by the attorney to perform experiments to prove or disprove an issue.

Qualifying as an Expert Witness

Receiving a subpoena to testify in a fire litigation does not give the fire investigator a carte blanche invitation to testify as an expert witness. To qualify, the fire investigator, in all probability, will be confronted with a barrage of questions concerning the witness's education, employment history, job classification, assigned responsibilities and any other subject matter related to his or her career. Although not difficult, the questions can be intimidating and can produce the feeling of vulnerability. To ease the dilemma of qualifying, the fire investigator should become thoroughly familiar with the proper functions of an expert witness.

There are two main types of witnesses who can testify in a court of law, the lay witness and the expert witness. A lay witness can only testify to the facts that he or she knows about the case. Conclusions based on those facts cannot be made by the lay witness. The conclusion will be interpreted by the trier of fact based on the testimony of the lay witness. Example: A lay witness can testify that he or she observed a red metal can with the word gasoline written on the side, setting on the living room floor. The lay witness, however, cannot conclude that the fire is an arson based on the observed facts.

The expert witness is an individual who has knowledge and expertise not found in ordinary people. Through this knowledge and expertise, the expert witness can give opinions and inferences from the facts which the trier of fact would not be competent to interpret. If the trier of fact can come to a decision without the assistance of an expert witness, then the testimony of the expert witness can be ruled inadmissible. In his article, *The Expert Witness on Criminal Cases*, Melvin Lewis states:

> It necessarily follows that no one may testify as a witness until it first appears, either from his own testimony or from some other source, that he is indeed

specially qualified to make judgements of which the trier of fact is incapable – or at least that he can make them more reliable than can the trier of fact.

The primary element for expert testimony to be admissible centers around Rule 702 of the Federal Rules of Evidence. Rule 702 states:

> If scientific, technical or other specialized knowledge will assist the trier of fact to understand the evidence or to determine a fact in issue, a witness qualified as an expert by knowledge, skill, experience, training or education may testify thereto in the form of an opinion or otherwise.

Based on Rule 702, the courts have generally agreed that two conditions must be met before a witness can qualify as an expert:

1. The witness must show through questioning that he or she is qualified through education, knowledge, training, skill, or experience to speak on the subject.
2. The knowledge of the witness is above the knowledge or understanding of the ordinary lay person and the knowledge given by the witness will be an aid to the understanding of the facts in the case.

There are no standards set on the minimum education, training, and experience needed to qualify. Each case will have a judge who will determine the extent of the qualifications needed to testify as an expert. This statement also brings up a very good point. Qualifying as an expert witness in one case does not automatically qualify the same witness as an expert in future cases. Every witness from the top member in his or her field to the novice must, when testifying each time as an expert witness, go through this process of qualifying.

The best way for the witness to organize all of his or her credentials is through a resume, also known as a curriculum vitae (Figure 15-1). If presented properly, this document can be an impressive way to present oneself to the prosecutor or opposing attorney. One should never lie on a resume. A good attorney will follow up on the validity of the credentials in an attempt to discredit the witness. This document should be organized into neat, typewritten categories of employment, education, training, certifications, and experiences. Always make sure the curriculum vitae is up to date. There are many ways a resume can be formatted. Figure 15.1 is just one example of how a resume can be presented.

AL E. GATOR
111 MAIN STREET
ANYPLACE, FLORIDA 33000
(407) 000 0000

EMPLOYMENT:

1992 to present — Chief Fire Investigator and Inspector, Bureau of Fire Prevention, City of Anyplace, Florida Fire Department.

1988 to 1992 — Firefighter/Paramedic, City of Anyplace, Florida Fire Department.

GENERAL EDUCATION:

- Attending P.B. Community College for a AA degree in Criminal Justice.
- Graduated from P.B. Community College with an AS degree in Fire Science.
- Graduated from Anyplace High School, Anyplace, Florida in 1986.

TRAINING:

- February, 1994 — Legal Issue Course, Florida State Fire College (40 hours)
- February, 1994 — Chemistry for Arson Investigators Course, Florida State Fire College (40 hours)
- March, 1994 — Cause and Origin Course, Florida State Fire College (40 hours)
- March 1994 — Latent Investigation Course, Florida State Fire College (40 hours)
- September, 1993 — 49th Florida Arson Seminar (40 hours)
- February, 1993 — Arson for Profit School, Bureau of Alcohol Tobacco and Firearms, (40 hours)
- August, 1992 — Graduated from the Police Arson School, sponsored by the Federal Bureau of Investigations (FBI), Quantico, VA. (40 hours)
- June, 1992 — Graduated from the National Fire Academy in Fire/Arson Investigation (120 hours)
- April, 1992 — Certificate of Competency in Fire Investigation by the State Fire Marshal's Office, Bureau of Standards and Training (24 hours)
- January, 1992 — Certificate of Competency in Arson Investigation by the State Fire Marshal's Office, Bureau of Standards and Training (24 hours)
- November 1991 — Certified as a Fire Inspector by the State Fire Marshal's Office, Bureau of Standards and Training (160 hours)
- May, 1990 — Rescue/Extrication Course, State Fire Marshal's Office, Bureau of Standards and Training (80 hours)
- April, 1989 — Certified Paramedic in the State of Florida, registry no. 4371
- October, 1988 — Certified Firefighter through the State Fire College, Ocala, Florida

EXPERIENCE:

- As of September 1, 1994 I have participated in 50 different practical fire scene burns to observe fire behavior and study the patterns of fire travel. Assorted incendiary devices have been also used to study the components of the device and to examine the debris after ignition.

- Testified 4 different times as an expert witness in the field of fire investigation and to the origin and cause of fires in the United States Federal Court.

◆ Testified as an expert witness in the field of fire investigation and to the origin and cause of fires in the 15th and 17th Florida Judicial Circuits for a total of 5 different times.

◆ As of September 1, 1994 I have investigated approximately 250 fires of which 125 were determined to be incendiary, 100 accidental and 25 undetermined.

PERSONAL INFORMATION:

◆ Current instructor at the P.B. Community College Fire and Law Enforcement Academy.
◆ Guest speaker at the IAAI 43rd Annual Seminar, May, 1992.
◆ Guest speaker at the Florida Advisory Committee on Arson Prevention's 16th Annual Seminar, November 1992.
◆ Current member since 1988 in the State chapter and the International Association of Arson Investigators.
◆ Current member of the National Fire and Arson Report.
◆ Organizer and member of the Arson Intelligence Force in Anyplace County.
◆ Co-organizer of the Anyplace County Fire-Arson Task Force.
◆ Have written several published articles relating to the field of Fire/Arson Investigations in the State and International IAAI Magazine.

Figure 15-1. A resume' of a fire investigator.

Since this book is written for the fire investigator, let's take some time to discuss the fire investigator qualifying as an expert witness. It is important to remember that the witness, in order to qualify, is going to give a brief history of his or her career as a fire investigator.

The attorney wishing to have the witness testifying as an expert will ask questions illustrating the expert's expertise. The questions will begin by asking the witness to state his or her name and the department, agency, or company he or she work for. Answers to the questions concerning employment should begin with the investigator's current position, the amount of time in that position, and the amount of time employed by the current employer. The investigator should briefly give a job description and duties assigned to the current position held. The attorney may further ask the investigator about prior employers if it can assist the witness in qualifying.

Education and training are probably the two most important factors taken into consideration by the judge to qualify an expert witness. Education is self-explanatory. Begin with high school and continue, if applicable, with college, the type of degree earned and any other advanced college degrees.

Training is another form of education that establishes the investigator in continually learning about the skills of his or her occupation.

When the fire investigator is asked about training, the witness should initiate the answer from the incipient stages of the career. For most investigators, this stage began as recruits in a fire or police academy. If so, commence from this point and conclude with the most recent training program attended. In this situation, the fire investigator should list as many special schools, courses, and seminars related to fire science and fire investigation as possible.

Expertise can be acquired through experience on the job. Many fire investigators from fire departments have been known to qualify as experts based on the number of years on the department in fire suppression activities. Proficiency can be obtained by documenting hands on experience on practical burns used to observe fire behavior and to conduct fire scene training exercises. Maintaining an up-to-date record of the number of fires investigated and their classifications benefits the witness in qualifying. Another record worth preserving and given in testimony is the number of times the fire investigator has previously testified as an expert witness as to the origin and cause of fires. Also include the names of federal and/or state judicial circuit(s).

Beyond education, training, and experience, other criteria such as the publication of articles in the field of expertise, membership in professional organizations, and teaching positions can promote the witness as an qualified expert.

The following testimony is a facsimile of the typical questions and answers asked of the witness who is qualifying as an expert witness.
Attorney (**A**): "Sir, could you please give the court your full name and spelling?"
Witness (**W**): "Yes, my name is Al E. Gator, A L E. G A T O R."
A: "Mr. Gator, for whom do you work?"
W: "I work for the City of Anyplace Fire Department."
A: "And how long have you worked for the City of Anyplace Fire Department?"
W: "Since 1988 or six years."
A: "What have been your duties with the Anyplace Fire Department?"
W: "I began my career with the Anyplace Fire Department as a firefighter. The city sponsored me at the Florida State Fire College to be state certified as a firefighter. Upon completion at the Fire College, I was assigned to the Combat Division, Engine 2 of Fire Station Number 1. After completing my probation period, I became a fire/paramedic

and was assigned to Rescue 2. In 1992, I was reassigned to the Bureau of Fire Prevention as a Fire Inspector. My duties at that time were to inspect buildings for Life Safety Code violations and fire investigations. In 1993, I became the Chief Fire Investigator for the Bureau."

A: "As Chief Fire Investigator, what were your duties?"

W: "To investigate all fires within the city limits and to determine the origin and cause of the fire. To help establish the cause of a fire, debris is examined, physical evidence collected and witness statements are documented. Once a fire is determined to be incendiary the city police department is notified and a combined investigation between the two departments is conducted. The follow-up investigation of an incendiary fire may include interrogating suspects, making arrests, and presenting the case to the state's attorney's office for prosecution and possible trial."

A: "Mr. Gator, as the Chief Fire Investigator for the city, have you had any training in the area of fire investigation?"

W: "Yes sir, I have."

A: "Describe your training."

W: "I attended a four-week course (160 hours) at the State of Florida Fire College in the field of Fire Investigation. Included in that course were legal Issues, determining the origin and cause of the fire, chemistry for fire investigators, and latent investigation. At the completion of the course, a test was given, at which time I passed and was awarded a Certificate of Completion.

I have graduated from the FBI Academy in a course entitled "Police Arson School." I graduated from the National Fire Academy in a three-week course in fire/arson investigations, and graduated from the ATF school in Georgia in the course entitled "Arson for Profit."

I have attended seminars in the field of fire/arson investigations within the state of Florida. I have taken several courses from the State Fire Marshal's Office, Bureau of Standards and Training in the field of Fire/Arson Investigations.

Prior to my assignment in the bureau I became a certified firefighter 1988."

A: "How many fires have you personally investigated as the fire investigator for Anyplace Fire Department?"

W: "As of September 1, 1994, I have investigated approximately 250 fires."

A: "How many of those fire were accidental?"

W: "One hundred."

A: "How many were incendiary?"

W: "One hundred twenty five.

A: "Do you belong to any organizations related to fire investigations?"

W: "Yes, I belong to the State Chapter and the International Association of Arson Investigators (IAAI), and I am a member of the National Fire and Arson Report.

I helped organize and am currently a member of the Anyplace County Task Force.

I am also an instructor at the P. B. Community College Fire and Law Enforcement Academy where I teach fire and criminal courses to recruits and degree seeking students."

A: "Have you written any articles or books in the field of fire investigation?"

W: "Yes, I have written an article for the 17th Judicial State Attorney's monthly periodical entitled, *The Role of Law Enforcement Personnel at Fire Scenes.* I also wrote an article for the IAAI bi-monthly periodical entitled, *Tunnel Vision is NOT A Tool For The Fire Investigator.*"

A: "What type of education do you have?"

W: "I graduated from Anyplace High School in 1986. Received an AS Degree in Fire Science and I am now currently attending P.B. Community College for an AA degree in Criminal Justice."

A: "Investigator Gator, of the various classes you have attended, have they included the training of determining the origin and cause of the fire?"

W: "Yes sir, they have."

A: "Have these classes included determining various fire causes?"

W: "Yes sir, they have."

A: "What is your definition of a fire?"

W: "Fire is the rapid oxidation of a substance accompanied by heat and light."

A: "What are the elements needed for a fire to occur?"

W: "Heat, fuel, oxygen, and a chain reaction."

A: "Investigator Gator, can you name the different ways heat can be transferred?"

W: "Yes sir. They are by convection, conduction and radiation."

A: "Please briefly describe each one."

W: "Convection is the transfer of heat by means of moving gases or liquid. Conduction is the transfer through a solid. And radiation is heat transfer from one material to another by heat energy waves."

A: "How does a fire investigator make a determination into the fire cause?"

W: "A fire investigator will determine a fire by observing fire indicators that will show burn patterns, fire travel, smoke patterns, and sometimes low burn or pour patterns and through a process of elimination eliminate all fire causes except for one."

A: "What is a flashover?"

W: "A flashover is a condition found in an enclosed space where all the ordinary combustible materials reach ignition temperature at the same time and spreads rapidly throughout the area."

A: "What is a back draft?"

W: "A back draft is a condition found within an enclosed area where combustion has taken place for a period of time. The oxygen levels within the space are low and the super heated fire is starving for oxygen. When a door is opened or window broken a rush of oxygen enters the room causing a vapor like explosion."

The list of questions can continue until the attorney feels the witness has established him or herself as an expert. If the witness has satisfactorily laid a foundation to qualify as an expert, the attorney will request from the court that the witness be considered an expert and entitled to give opinions. Before granting the request, the court will give the opposing attorney the opportunity to ask questions (voir dire) of the witness in an attempt to break the established foundation.

In this particular case, the opposing attorney reviewed the witness's resume prior to the trial and decided to stipulate on the qualifications. Further questioning would only enhance the creditability of the witness. The tactical approach of the adversary attorney could center on the chance to discredit or weaken the testimony of the witness during cross-examination. This maneuver would cause greater damage than voir dire.

Therefore, Investigator Gator has been allowed to testify as an expert witness. However, before continuing with the testimony, the judge has ordered a ten-minute recess. This short break will enable the reader to learn more about the expert witness.

Before expert opinion can be given concerning the origin and cause of a fire, the testimony must be based on the facts observed by the witness and not a guess or speculation. The expert witness can communicate expert opinion in one of two ways: through firsthand knowledge of the facts, or, give an opinion based on certain facts already established through other witnesses.

The majority of the expert opinions will be through firsthand knowledge. Testimony of the witness will encompass a detailed explanation of how the fire investigation was conducted from beginning to end in a chronological order. The witness, through field notes and the use of demonstrative evidence, such as photographs, sketches, and physical evidence, will describe:

- How the fire scene was discovered.
- The investigator's observations.
- How the least area of damage brought the investigator to the area of origin.
- How burn patterns, knowledge of fire science, fire load, the elimination of ignition sources except for one, and, any other information that was utilized to determine the origin and cause of the fire.

The majority of fire investigators, not having firsthand knowledge of the scene, will be experts for the opposing attorney. These investigators will probably process a fire scene already disturbed by the public sectors' investigation, or review all reports, photographs, sketches, etc. from the initial fire investigator. After the initial investigator gives his or her testimony, the adversary attorney will use their expert in a rebuttal as to the origin and/or cause of the fire. By using the facts and evidence introduced by the initial fire investigator, the adversary witness may testify, if qualified, as an expert and come up with a different conclusion.

The opinion of any fire investigator expert witness is determined by the trier of fact according to the same tests applied in the evaluation of ordinary evidence. The trier of fact is not bound to accept opinions and may reject any opinion they feel is unreasonable or unwarranted.

The judge has returned to the courtroom and Investigator Gator is about to continue:

A: "Investigator Gator, on the date of the fire, were you employed by the City of Anyplace Fire Department?"

W: "Yes sir, I was."

A: "How were you notified about the fire located at _____?"

W: "I received a telephone from dispatch at approximately 2:00 AM."

A: "What did dispatch tell you?"

W: "That there was a structure fire and Chief Bono requested a fire investigator to respond to the scene."

A: "Did you go to the fire scene?"

W: "Yes sir, I did."

A: "How long did it take you to get to the fire scene?"

W: "Approximately one hour."

A: "From the initial time the fire department received the call and you arrived on scene, how much time had elapsed?"

W: "Approximately 2 1/2 hours."

A: "Upon arrival what did you do?"

W: "I went to the Incident Command Post to check in and to find the officer in charge."

A: "Then what did you do?"

W: "I took a quick walk around the perimeter of the structure to examine the exterior."

A: "Then what did you do?"

W: "I took statements from the first-in firefighters and witnesses."

A: "Then what did you do?"

W: "I began the fire scene examination by starting from the northwest exterior corner of the structure and by working in a clockwise direction examined and documented the exterior exposures."

Without getting into detail, the next series of questions dealt with Investigator Gator's observations and documentation of the fire scene.

A: "Investigator Gator, based on your knowledge as a fire investigator and your examination of the fire scene at *location of the fire scene* on *date and time of the investigation,* have you formed an opinion as to the area of origin of the fire?"

W: "Yes sir, I have."

From this point on, Investigator Gator will give a detailed description of the indicators used to determine the area of origin, point of origin, and fire cause. Within this part of the testimony, demonstrative evidence can be used to help bring the scene into the courtroom and aid the trier of fact in understanding the testimony.

Hopefully, the information in this book will lessen the chances of a poor performance of a fire investigator in the courtroom. Through the knowledge of how to enter, document, and present the fire scene and investigation, the fire investigator can improve the odds in his or her favor.

Do not be discouraged by mistakes. Never accept the mistake as a failure. Dissect and analyze any problems that may exist on the fire scene or in the courtroom. Through recognized oversights, omissions, slips, and stumbles, learn to improve techniques and to build confidence. Confucius once said, "Learning without thought is labor lost; thought without learning is perilous."

Fire safely yours,

Wayne P. Petrovich

VIDEO BIBLIOGRAPHY

Adams, Rich. Ems Videotape Training. *Firehouse.* April, 1992: 14

Allen, Diane M. Admission of Visual Recordings of Events or Matter Giving Rise to Litigation or Prosecution. *American Law Report.* 4th ed. Vol. 41. Rochester, NY: The Lawyers Cooperative Publishing Co. 1985: p. 812

Allen, Diane M. Admissibility of Visual Recording of Event or Matter Other Than That Giving Rise To Litigation or Prosecution. *American Law Reports.* 4th ed. Vol. 41. Rochester, NY: The Lawyers Cooperative Publishing Co. 1985: p.877.

Belli, Melvin M., Sr. *Modern Trials.* 2nd. ed. Vol. IV. St. Paul, MN: West Publishing Co. 1982.

Buying Your First Camcorder. *Popular Photography.* November, 1991: 77-78

Cheshire, David. *The Book of Video Photography.* London, England: Dorling Kindersley Limited., 1990.

Drechsler, Carl T. Admission of Videotape Film in Evidence in Criminal Trial. *American Law Reports.* 3rd ed. Vol. 60. Rochester, NY: The Lawyers Cooperative Publishing Co., 1974: p. 333.

The Enlightened Video Photographer. *Popular Photography.* November, 1991: 86-89

Evidence–as Motion Pictures. *American Jurisprudence, Proof of Facts.* Vol. 8. Rochester, NY: The lawyers Cooperative Publishing Co., 1960: p. 153.

Evidence–as Motion Pictures. *American Law Reports.* 2nd ed. Vol. 62. Rochester, NY: The Lawyers Cooperative Publishing Co. 1958: p. 686.

Evidence–Photographs/Motion Pictures. *American Jurisprudence.* 2nd ed. Vol. 29. Rochester, NY: The Lawyers Cooperative Publishing Co., 1967.

Evidence–Use of Motion Pictures. *American Law Reports-Later Case Service.* Vol. 62 66. Rochester, NY: The Lawyers Cooperative Publishing Co., 1984.

Gruber, Jordan S. Foundation for Contemporaneous Videotape Evidence. *American Jurisprudence–Proof of Facts.* 3rd ed. Vol. 16. Rochester, NY: The Lawyers Cooperative Publishing Co. 1992: p. 493.

Gruber, Jordan S. Videotape Evidence. *American Jurisprudence Trials.* Vol. 44. Rochester, NY: The Lawyers Cooperative Publishing Co., 1992.

Galluzzo, Tony. Essential Guide To Video. *Popular Photography.* November, 1991: 53-54

Joseph, Gregory P. Demonstrative Videotape Evidence. *Trials.* June, 1986: 60-66.

Lewis, Roland. *Home Video Maker's Handbook.* New York: Crown Publishers, Inc., 1987.

Norris, R.C. *The Complete Handbook of Super 8 Film Making.* Blue Ridge Summit, PA: Tab Books, 1982.

Quinn, Gerald V. *The Camcorder Handbook.* Blue Ridge Summit, PA: Tab Books Inc. 1987.

Robinson, Richard. *The Video Primer.* New York: Perigee Books, Putnam Publishing Group, 1983.

Scott, Charles C. Photographic Evidence. 2nd ed. Vol. III. St. Paul MN: West

Publishing Co., 1969.

Shake, Rattle, and Roll: How to Prevent It. *Popular Photography.* May, 1991: 54

Should You Upgrade Your Present Camcorder? *Popular Photography.* November, 1991: 80-81.

Techniques: Shooting It the Right Way. *Popular Photography.* November, 1991: 84-85

Thomas, Erwin Kenneth, *Make Better Videos with Your Camcorder.* Blue Ridge Summit, PA: Tab Books, Inc., 1991.

Traister, Robert J. *Make Your Own Professional Home Video Recordings.* Blue Ridge Summit Ridge, PA. Tab Books, Inc., 1982.

Video Cares and Caution. *Popular Photography.* November, 1991: 83

Videotape: What To Look For. *Popular Photography.* November, 1991: 88

SKETCH BIBLIOGRAPHY

Arson Investigation. Sacramento, CA: California District Attorney's Association.

Ballard, Scott T. *How to Be Your Own Architect.* White Hall, VA: Betterway Publications, 1987.

Brannigan, Francis L., Richard G. Bright & Nora H. Jason. *Fire Investigation Handbook.* Washington D.C: National Bureau of Standards Handbook 134, United States Printing Office, 1980.

Criminal Investigations. 2nd. ed. Gaithersburg, MD: International Association of Chiefs of Police, 1971.

Drawing To Illustrate Testimony. *American Law Reports: Annotated.* 2nd. ed. Vol. 9. Rochester, NY: The lawyers Cooperative Publishing Co. 1950 p. 1044.

Evidence. *American Jurisprudence.* 2nd ed., Vol. 29 A. Rochester, NY: The Lawyers Cooperative Publishing Co., 1994.

Fire and Explosion Investigations: Pamphlet 921. Quincy. MA: National Fire Protection Association, 1992.

Gard, Spencer A. *Jones on Evidence-Civil and Criminal.* Vol. III. Rochester, NY: The Lawyers Cooperative Publishing Co., 1972.

Helper, Donald E. & Paul Wallach. *Architecture Drafting and Design.* New York: Webster Div. McGraw-Hill Book Co., 1965.

Hergan, John J. *Criminal Investigation.* New York: McGraw-Hill Book Co., 1974.

Kennedy, John. *Fire, Arson and Explosion Investigation.* Chicago, IL: Investigations Institute.

Maps, Diagrams, and Models. *American Jurisprudence: Proof of Facts. Vol. 7.* Rochester, NY: The Lawyers Cooperative Publishing Co., 1960, p.601.

Muller, Edward J. *Reading Architectural Working Drawings. Vol. 1.* Englewood Cliffs, NJ: Prentice Hall, 1988.

Purver, Jonathan M. Foundation for Admission of Map, Diagram, or Sketch. *American Jurisprudence: Proof of Facts.* 2d. ed. Vol. 44 Rochester, NY: The Lawyers Cooperative Publishing Co., 1986.

Strong, John William. Demonstrative Evidence. *McCormick on Evidence.* St. Paul MN: West Publishing Co., 1992.

Wigmore, John Henry. *Evidence in Trial at Common Law. Vol. III.* Revised by James H. Chadbourn. Boston, MN: Little, Brown and Co., 1970.

Winebrenner, R. *Sketching Techniques.* West Palm Beach Police Department, 1982.

Wolverton, Mike & Ruth Wolverton. *Draw Your Own House Plans.* Blue Ridge Summit, PA. Tab Books, Inc, 1983.

PHOTOGRAPHY BIBLIOGRAPHY

Bailey, Adrian & Adrian Holloway. *The Book of Color Photography.* New York: Alfred A. Knopf, 1979.

Blaker, Alfred A. *Photography Art and Technique.* San Francisco, CA: W.H. Freeman, 1980.

Davis, Phil. *Photography.* New York NY: Mayflower Books, Inc., 1979.

Evidence. *American Jurisprudence* 2nd ed. Vol. 29 A. Rochester, NY: The Lawyers Cooperative Publishing Co., 1994.

Feller, Howard J. Photographic Evidence: More Than Meets the Eye. *Maine Bar Journal* Vol. 8, No. 6 Nov. 1993: 372-374.

Film, History of, *Academic American Encyclopedia.* Grolier Electronic Publishing, Inc., 1992.

Fire & Arson Photography. Eastman Kodak Company, 1977.

Gard, Spencer A. *Jones on Evidence–Civil and Criminal.* Vol. III. Rochester, NY: The Lawyers Cooperative Publishing Co., 1972.

Hayes, Paul W. & Scott M. Worton. *Essentials of Photography.* Indianapolis, IN: Bobbs-Merrill Company Inc., 1983.

The Joy of Photography. Eastman Kodak Company, 1979.

Keppler, Herbert & Kenny Yamamoto. How to Care for Your Camera. *Popular Photography.* Oct. 1991.

Lyons, Paul Roberts. *Techniques of Fire Photography.* Boston MA: NFPA (National Fire Protection Agency), 1978.

Patton, Murray. Flashlight Provides Lighting for Arson Photography. *Fire Chief Magazine,* June, 1982 42-44.

Peige, John D. *Photography for the Fire Service.* Oklahoma State University: Fire Protection Publications, 1977.

Photographs as Evidence. *American Jurisprudence : Proof of Facts.* Vol. 9. Rochester, NY: The Lawyers Cooperative Publishing Co., 1961.

Redsicker, David R. *The Practical Methodology of Forensic Photography.* New York: Elsevier Science Publishing Co., 1991.

Scott, Charles C. *Photographic Evidence Preparation and Presentation,* 2nd ed. Vols. 1, 2, 3 and Supplement, St. Paul MN: West Publishing Co., 1991.

Sussman, Aaron, *The Amateur Photographer's Handbook,* 8th ed. New York: Thomas Y. Crowall Co., 1973.

Using Photography to Preserve Evidence. Eastman Kodak Co., 1976.

Wigmore, John Henry. *Evidence in Trial at Common Law*. Vol. III. Revised by James H. Chadbourn. Boston, MA: Little, Brown and Co., 1970.

MODEL BUILDING BIBLIOGRAPHY

Belli, Melvin M. Sr. *Modern Trials*, 2nd ed., Vol. 4. St. Paul, MN: West Publishing Co., 1982.

Evidence. *American Jurispurdence*, 2nd. ed., Vol. 29 A. Rochester, NY: The Lawyers Cooperative Publishing Co., 1994.

Evidence: Model of Object or Site. *American Law Reports*, 2nd ed., Vol. 69. Rochester, NY: The Lawyers Co-Operative Publishing Co. 1960: p.424.

Hohauser, Sanford. *Architectural and Interior Models*. New York: Van Nostrand Reinhold, 1970.

Kennelly, John J. Use of Demonstrative Evidence, Including Models. *Trial Lawyer's Guide*. Vol. 16, No. 4, Winter 1973: 417.

Kicklighter, Clois E. *Architecture Residential Drawing and Design*. South Holland, IL: The Goodheart-Willcox Co. Inc. 1990.

Knoll, Wolfgang & Martin Hechinger. *Architectural Models: Construction Techniques*. New York: McGraw-Hill, Inc., 1992.

Koepke, Marguerite L. *Model Graphics: Building and Using Study Models*. New York: Van Nostrand Reinhold, 1988.

Maps, Diagrams, and Models. *American Jurisprudence, Proof of Facts*. Vol. 7. Rochester, NY: The Lawyers Cooperative Publishing Co., 1960. Supplement 1994.

Moore, Fuller. *Model Builder's Notebook*. New York: McGraw-Hill Publishing Co., 1990.

Photographs, Maps, Diagrams, Drawings, and Models. *American Jurisprudence*, 2nd ed. Vol. 29. Rochester, NY: The Lawyers Cooperative Publishing Co., 1967.

Preparing and Using Models. *American Jurisprudence Trials*. Vol.3 Rochester, NY: Lawyers Cooperative Publishing Co., 1989 p. 377.

Wesolowski, Mary & Wayne. *Modell Railroad Structures From A to Z*. Newton N.J: Carstens Publication, Inc., 1984.

Working With Wood. *Time Life Books*. Alexandria, VA: Time Magazine, 1979.

LAW BIBLIOGRAPHY

American Jurisprudence. 2nd ed., Vol. 29, Rochester, NY: The Lawyers Cooperative Publishing Co., 1967.

American Law Reports. 2nd ed. Vol. 31. Search and Seizure–Consent. Rochester, NY: The Lawyers Cooperative Publishing Co., 1958: 1078.

American Law Reports. 4th ed. Rochester, NY: The Lawyers Publishing Co., 1985.

Coughlin, George Gorden. *Your Introduction to Law.* New York: Barnes & Noble Books, 1975.

Dripps, Donald A. It Might Be a Roust, But It Isn't a Seizure. *Trial.* Vol. 28 No. 1 January, 1992: 66-68.

Fave, Wayne R. *Search & Seizure: A Treatise on the Fourth Amendment.* Vols. I & III, St. Paul, MN: West Publishing Co., 1987.

Ferdico, John N. *Criminal Procedure for the Law Enforcement Officer.* St. Paul MN: West Publishing Co., 1980.

The Living Law. 1987-1988 A Guide to Modern Legal Research. Rochester, NY: The Lawyers Cooperative Publishing Co., 1988.

Ludington, John P. Search and Seizure–What Constitutes Abandonment of Personal Property Within the Rule that Search and Seizure of Abandoned Property Is Not Unreasonable. *American Law Reports.* 4th ed., Vol. 40. Rochester, NY: The Lawyers Cooperative Publishing Co., 1985: 381.

Kruk, Theresa Ludwig. Admissibility in Criminal Case of Evidence Discovered by Warrantless Search in Connection with Fire Investigation–Post Tyler Cases. American Law Reports 4th ed. Vol. 31. Rochester, NY: The Lawyers Cooperative Publishing Co., p. 194.

Sample Pages. 3rd ed. St. Paul MN: West Publishing Co., 1986.

Search and Seizure–Consent. *American Law Reports.* 2nd ed., Vol. 31. Rochester, NY: The Lawyers Cooperative Publishing Co., 1953: 1078.

Waltz, Jon R. *Criminal Evidence.* Chicago, IL: Nelson-Hall Co., 1983.

TESTIFYING BIBLIOGRAPHY

Baldwin, Scott. Cross Examination. *Trial.* July, 1987: 76-81.

Beering, Peter S. How to Make Love to a Prosecutor or How Not to Get Screwed by a Prosecutor. Indianapolis/Marion County Prosecutor's Office, 1992.

Beering, Peter S. Secrets To A Successful Arson Prosecution. *Firehouse.* Sept. 1989: 69.

Beering, Peter S. You Are Invited...To Be a Witness. *Firehouse.* August 1990: 41.

Belli, Melvin M. Sr. *Modern Trials.* 2nd ed., Vol. III. St Paul MN: West Publishing Co., 1982.

Berry, Dennis J. *Fire Litigation Handbook.* Quincy MA: National Fire Protection Association, 1984.

Burnette, Guy E. Jr. Expert Testimony in an Arson Case. *Informer.* (Florida Chapter of the International Assosication of Arson Investigators), Summer, 1993: 17-27.

Burnette, Guy E. Jr. Legal Issues in Fire Investigations. Presented during Arson Investigation Program at the Florida State Fire College, Ocala, Fl. March 14, 1991.

Burns, James Mac Gregor, J.W. Peltason & Thomas E. Cronin. *Government by The*

People. 14th ed. Englewood Cliffs, NJ: Prentice-Hall, 1990.

Casamassima, Thomas J. & David A. Borehert. The Expanding Discovery of Expert Opinions and Reports. *For the Defense.* Vol. 28 No. 4, April 1986: 15-23.

Dunn, R. Roog. Persuasive Testimony. *Informer.* (The Florida Chapter of the International Association of Arson Investigators), Summer 1993: 49-57.

Ekman, Paul. *Telling Lies.* New York: W.W. Norton & Co., 1992.

Ferdico, John N. *Criminal Procedure for the Law Enforcement Officer.* St. Paul, MN: West Publishing Co., 1980.

Langerman Amy G. Making Sure Your Experts Shine–Effective Presentation of Expert Witness. *Trial.* Vol. 28 No. 1, Jan. 1992: 106-110.

Lewis, Melvin B. The Expert Witness in Criminal Cases. *Criminal Defense.* Vol. 3, No. 1, Jan. 1976.

Moore, Thomas A. Cross-Examining the Defense Expert. *Trial.* Vol. 27 No. 5, May, 1991: 49-52.

O'Neil, John H. Jr. Have I Got a Case for You: Surviving Your First Jury Trial. *Maine Bar Journal.* Vol. 8 No. 6, November, 1993: 364-367.

Somers, Clifford L. Deposing an Adverse Witness. *For the Defense.* Vol. 31 No. 7, July, 1989.

Strong, John William. *McCormick on Evidence.* 4th ed., Vol. 1, St. Paul, MN: West Publishing Co., 1992.

Tilton, Dennis. *Taking The Stand.* North Hollywood, CA: Film Communications, 1981.

Torcia, Charles E. *Wharton's Criminal Evidence.* Vol. III. Rochester, NY: The Lawyers Cooperative Publishing Co., 1987.

The Use and Misuse of Expert Evidence in the Courts. Judicature–*The Journal of the American Judicature Society.* Vol. 77 No. 2, September-October, 1993: 68-76.

Wigmore, John Henry. *Evidence in Trial at Common Law.* Vol. V. Revised by James H. Chadbourn. Boston, MA: Little, Brown and Co., 1974.

COMPUTER BIBLIOGRAPHY

Approaching Arson Investigations Armed with Facts, Science and a Computer. *Firehouse.* February, 1994: 50-52.

Architects Give Video "Walk Through" Tour of Buildings. *Presentation for the Visual Communication.* July, 1993: 24-25.

Beyler, Craig. Introduction to Fire Modeling. *Fire Protection Handbook.* 17th ed. Quincy, MA: National Fire Protection Association, 1991: 10-82 thru 10-85.

Christen, Hank. The Computer World of 1994. *Firehouse.* February, 1994: 26-28.

Covington, Michael & Douglas Downing. Dictionary of Computer Terms. 3rd ed. Hauppauge, NY: Barron's Educational Series, Inc. 1992.

Di Nenno, Philip J. The Future of Fire Modeling. *Fire Protection Handbook.* 17th ed. Quincy, MA: National Fire Protection Association, 1991: 10-124 thru 10-128.

Jasanoff, Sheila. What Judges Should Know About the Sociology of Science.

Judicature–The Journal of the American Judicature Society. Sept-Oct. 1993: 77-82.

Kelso, Clark J. Judicial Technology in the Courts. *American Jurisprudence Trials.* Vol. 44. Rochester, NY: The Lawyers Publishing Cooperative Publishing Co., 1992.

Lehr, Louis A. Jr. Admissibility of a Computer Simulation. *For The Defense.* July, 1990: 8-11.

Marcotte, Paul. Animated Evidence: Delta 191 Crash Re-Created Through Computer Simulations at Trial. *American Bar Association.* December, 1989.

Martin, E.X. III. Using Computer-Generated Demonstrative Evidence. *Trial.* September, 1994: 84-88.

The National Center For State Courts [NCSC] Technology Programs. *The Judge's Journal.* Vol. 32 No. 3, Summer 1993: 74-79.

Nelson, Harold E. Application of Fire Growth Models to Fire Protection Problems. *Fire Protection Handbook.* 17th ed. Quincy, MA: National Fire Protection Association, 1991: 10-109.

O' Connor, Kathleen M. Computer Animations in the Courtroom: Get with the Program. *The Florida Bar Journal.* November, 1993: 20-28.

Ryan, Michael. Go Anywhere! But Don't Leave Your Chair. *Parade Magazine.* March 21, 1993: 18.

Schaefer, C. Caverhill, John L. Messina & R.J.H. Bollard. Computer Simulations in Court. *Trial.* July, 1987: 69-74.

Schroeder, Erica. 3-D Studio Gives Crime-Solving A New Twist. *PC Week Software.* March 9, 1992: 51.

Simmons, Robert & J. Daniel Lounsbery. Admissibility of Computer-Animated Reenactments in Federal Courts. *Trials.* Septmenber, 1994: 78-83.

Smith, Raoul. *The Facts on File Dictionary of Artificial Intelligence.* New York: Facts On File, 1989.

Spencer, Donald D. phD. *Computer Dictionary,* 3rd ed. Ormond Beach, FL: Camelot Publishing Co., 1992.

Walton, William D. & Edward K. Budnick. Deterministic Computer Fire Models. Fire Protection Handbook. 17th ed. Quincy, MA: National Fire Protection Association, 1991: 10-86 thru 10-91.

Watts, John M. Probabilistic Fire Models. *Fire Protection Handbook.* 17th ed. Quincy, MA: National Fire Protection Association, 1991: 10-93 thru 10-98.

Yule, John David. *Concise Encyclopedia of Science and Technology.* New York: Crescent Books, 1985.

DEMONSTRATIVE EVIDENCE BIBLIOGRAPHY

Bailey, William S. Demonstrative Evidence–Storyboards: Inexpensive and Effective. *Trials.* September, 1994: 64.

Brain, Robert D. & Daniel J. Broderick. Demonstrative Evidence–Clarifying Its Role at Trial. *Trial.* Septmeber, 1994: 73-76.

Gard, Spencer A. *Jones on Evidence–Civil and Criminal.* Vol. 3. Rochester, NY: The Lawyers Cooperative Publishing Co., 1972.

Givens, Richard A. *Demonstrative Evidence.* Colorado Springs, CO: Shepards/McGraw Hill, 1994.

Grimes, William A. *Criminal Law Outline.* Reno, NV: The National Judicial College, University of Nevada, 1991.

Heninger, Stephen D. Demonstrative Evidence–Cost-Effective Demonstrative Evidence. *Trials.* September, 1994. 65-66.

Strong, John William. *McCormick on Evidence.* Vol. II. St. Paul MN: West Publishing Co., 1992.

Tarantino, John A. *Trial Evidence Foundations.* Santa Ana, CA: James Publishing Co., 1990.

Torcia, Charles E. *Wharton's Criminal Evidence.* 13th ed. Vol. III Rochester, NY: The Lawyers Cooperative Publishing Co., 1973.

Turbak, Nancy J. Demonstrative Evidence–Accentuate the Positive. *Trials.* September, 1994: 63-64.

Wigmore, John Henry. Evidence in Trial at Common Law. Vol. III. Revised by James H. Chadbourn Vol. III. Boston, MA: Little, Brown and Co., 1970, p. 91.

INDEX

A

Abandonment, 51
Abel vs. United States, 55
Administrative search warrant, 13, 14, 18-28
Admissions, 66
Analytical fire models, 262
Appellate jurisdiciton, 272
Arizona vs. Hicks, 47
Arizona vs. Robertson, 74
Arkansas State Highway Commission vs.
    Rhodes, 207
Artifical intelligence, 263-265
Artuor Dominquez vs. Florida, 34, 35
ASA (American Standards Association), 134

B

Barham vs. Norell, 164
Baseline Method of measurement, 105
Beckwith vs. United States, 73
Berkemer ve. McCarthy, 74
Beverly Hills Fire Litigation, 230
Body language, 293-297
Brady vs. United States, 78
Bram vs. United States, 68
Break lines in sketching, 94
Brown vs. Mississippi, 68
Bumper vs. North Carolina, 30
Burden of proof, 277, 278
Burriss vs. Texaco, inc., 207
Butt joint, 241
Butt joint with a cover, 242

C

California vs. Ciraolo, 60
Camera vs. Municipal Court, 13, 19-21
Closing arguments, 282
Commonwealth vs. Roller, 171
Computer animation, 252
Computer simulation, 253

Computer sketch, 119
Computers, 249-265
    admissibility, 255-261
    artificial intelligence, 263-265
    expert systems, 264
    fire models, 261-263
Confessions, 66-71
Connecticut vs. Joyce, 52-54
Connecticut vs. Zindros, 55
Consent to search, 29-39
    accelerant detection canines, 34
Contemporaneous videotape, 191
Contour lines, 225
Coolidge vs. New Hampshire, 43-46
Corpus delicti, 278
Courts, 270-273
Court of appeals, 271
Cross examination, 281, 297-300
Culombe vs. Connecticut, 68
Curriculum vitae, 304-306
Curtilage, 59, 61-63
Custody, 74-77

D

Datum point, 225
Dauber vs. Merrell Dow Pharmaceutical,
    259
Davis vs. United States, 76
Demonstrative evidence, 83-87
Detailed sketch, 102
Dimension lines, 94
DIN (deutsche Industrie Norm), 134
Diorama, 214
Direct examination, 278
Discovery, 277
Dow Chemical Co. vs. United States, 63, 64
Duncan vs. Kieger, 171
Dunn vs. United States, 61-63

323

E

Edwards rule, 73
Edwards vs. Arizona, 73
Elevation sketch, 103
Ercoli vs. United States, 67
Escobedo vs. Illinois, 68, 69
Evidence, 279
Exclusionary rule, 5
Extigent or emergency circumstance, 9
Expert witness, 301-312
Exploded view or a sketch, 102
Extension lines, 94

F

Federal courts, 269-272
Finished sketch, 116
Floor plan, 101
Floor plan model, 211
Forte vs. United States, 67
Fourth amendment, 5
Frank vs. Maryland, 19
Fruit of the poison tree doctrine, 5
Frye Test - see Frye vs. United States
Frye vs. United States, 256-259

G

Gallick vs. Novotney, 206
Gerrard vs. Porcheddu, 229
Go-Bart Importing Co. vs. United States, 42
Green vs. City and County of Denver, 127
Grid method of measurement, 109
Gulombe vs. Connecticut, 68

H

Handford vs. Cole, 91
Harris vs. United States, 43, 44
Hassam vs. Safford Lumber Co., 206
Hendricks vs. Swenson, 197
Hester vs. United States, 51, 52
Hidden lines, 94
Hoffa vs. Unites States, 72
Hopt vs. Utah, 67
Houses, 7
Hudson vs. Texas, 54
Hung jury, 283

I

Immediately subsequent videotape, 192
Interogation, 71-74
Invoking the rule (see Rule of sequestration)
ISO (International Standards Organization), 135
Joints, in model construction, 240-244
Judicial system, 269-283

K

Katz vs. United States, 7

L

Lay witness, 303
Lopez vs. Texas, 198
Luco vs. United States, 126

M

Mapp vs. Ohio, 5
Marshall vs. Barlow's Inc., 22
McCotter Transport Co. vs. Hall, 91
Measurements in a sketch, 104-111
Michigan vs. Clifford, 15-19, 35, 46
Michigan vs. Tyler, 9-14
Milligan's Heirs vs. Hargrove, 89
Miranda vs. Arizona, 69, 70
Miranda Warnings, 33, 65-79
    waiver, 78
Models, 203-247
    admissibility, 204-209
    base of, 239
    construction of, 227
    latex paint, 236
    legend, 246
    materials for construction, 231-237
    painting the model, 236, 237
    site elements, 244
    tools, 238
    types of, 209-227
    work area, 238
Montana vs. Smith, 92
Motion (in the legal system), 277
Motion of immediate dismissal, 282
Motion to suppress, 277

N

Neighborhood model, 218
Neighborhood sketch, 99
New Jersey vs. Fox, 206
New York vs. Quarles, 72
Noe vs. Florida, 73
Non verbal communication (*see* Body
  language)

O

Object lines, 93
Oliver vs. United States, 58-60
Omnidirectional microphone, 177
Open fields, 57-64
Opening statements, 277
Opper vs. United States, 67
Oregon vs. Bradshaw, 74
Oregon vs. Mathiason, 73, 75
Original jurisdiction, 270, 272
Owensby vs. Indiana, 92

P

Paramore vs. Florida, 194
Pennsylvania vs. Muniz, 72
People vs. McHugh, 258
People vs. Riddle, 72
People vs. Speck, 207
Perma Research and Development vs.
  Singer Co., 255
Photo log, 137, 152
Photographs, 83, 125-167, 286
  admissibility of photographs, 164-167
  aerial photographs, 155
  Barham vs. Norell, 164
  daguerrian process, 126
  documentation of the fire scene, 136-151
  exterior photographs, 140
  film, 134-136
  history of, 125-128
  interior photographs, 140
  lighting, 155-162
  maintenance of the camera, 162, 163
  painting the light, 156-161
  photo log, 137, 152
  photographing a fire fatality, 154, 155
  photographing evidence, 154

  pictorial testimony, 165, 192
  silent witness, 166
  types of cameras, 130-134
Pictorial communication, 152
Pictorial fire models, 261
Plain View Doctrine, 41-49
Procedures of a trial, 277-279

R

Rebuttal, 282
Rebuttal witness, 282
Redirect examination, 281
Rectangular Method of measurement, 107
Reinke vs. Sanitary District, 205
Retrospective videotape, 192
Rhode Island vs. Innis, 71
Rogers vs. North Carolina, 92
Rough sketch, 104
Rule of Sequestration, 278

S

Schneckloth vs. Bustamonte, 29, 30
Searches, 5-64
See vs. Seattle, 13, 19-22
Shaeffer vs. General Motors Corp., 257
Shook vs. Pate, 89
Silent witness theory, 166, 192
Sketches, 83, 89-123
  computer sketch, 119
  finished sketch, 116-123
  line weights, 93, 94
  measurements within a search, 104-110
  rough sketch, 104-116
  types of, 99-103
  types of measurements, 105-111
South Carolina vs. Lemacks, 54
Splice joint, 243
Stansbury vs. California, 74
Starr vs. Campos, 257
State courts, 272
Steven M. Solomon Jr. vs. Edgar, 193
Stoner vs. California, 32
Stylization, 229
Subpoena, 273
Subpoena duces tecum, 275
Substantive evidence, 192
Supreme court, federal, 271

T

Texas vs. Brown, 45, 46
Topography model, 222
Tort, 278
Triangulation method of measurement, 108
Tucson vs. LaForge, 207

U

Udderzook vs. Pennsylvania, 126
Underwood vs. Indiana, 92
Unidirectional microphone, 176
United States vs. Booth, 71
United States vs. Brady, 72
United States vs. Feldman, 71
United States vs. Lee, 41
United States vs. Lefkowitz, 42, 43
United States vs. Marron, 41, 42
United States vs. Matlock, 31
United States vs. Taylor, 42
United States vs. Willis, 78

V

Video, 169-201
  admissibility, 193-199
  back light, 174
  close-up shot, 180
  composition, 179-181
  documenting a confession, 197
  documenting experiments, 198
  documenting reenactments, 198, 199
  fill light, 174
  fire fatality, 190
  high angle shot, 185
  key light, 174
  history, of, 170, 171
  lighting, 173-176
  low angle shot, 186
  lux, 173
  middle shot, 179
  panning, 187
  pictorial communication, 165, 192
  retrospective videotape, 192
  silent witness theory, 192, 196
  sound, 170-178
  three-point lighting, 174
  two point lighting, 174
  types of video cameras, 171-173
  zoom lens, 187
Virtual reality, 264
Voir dire, 277

W

W.R. Smith vs. Ohio Oil Co., 85, 205
Washington vs. Henry LeRoy Gray, 208
Weeks vs. United States, 5
William F. Sherman vs. City of Springfield,
  Illinois, 205
Wilson vs. United States, 165
Witness, 301-313
Wofford vs. Oklahoma, 91
Writ of certiorari, 272

Y

Yeagley vs. Indiana, 90